Science
and Cooking

Science
and Cooking

PHYSICS MEETS FOOD,
FROM HOMEMADE
TO HAUTE CUISINE

**Michael Brenner, Pia Sörensen,
and David Weitz**

W. W. NORTON & COMPANY
Independent Publishers Since 1923

For information about permission to reproduce selections from this book, write to
Permissions, W. W. Norton & Company, Inc., 500 Fifth Avenue, New York, NY 10110

For information about special discounts for bulk purchases, please contact
W. W. Norton Special Sales at specialsales@wwnorton.com or 800-233-4830

Manufacturing by TransContinental
Book design by Faceout Studio
Production manager: Julia Druskin

ISBN 978-0-393-63492-1

W. W. Norton & Company, Inc., 500 Fifth Avenue, New York, N.Y. 10110
www.wwnorton.com

W. W. Norton & Company Ltd., 15 Carlisle Street, London W1D 3BS

1 2 3 4 5 6 7 8 9 0

The discovery of a new dish confers more happiness on humanity
than the discovery of a new star.

—Jean-Anthelme Brillat-Savarin, *Physiologie du goût*, 1825

Contents

Preface

Ten years ago, when we first dreamed of a Science and Cooking course at Harvard University, I dared to predict that we were at a historic moment. A decade later, this project has enjoyed an enormously fruitful continuity, such that it is now reflected in the book you are currently reading. The initial idea for this interesting synergy was very clear: since cooking is something familiar to everyone, it is an ideal way of motivating students into learning physics concepts such as phase transitions, emulsions, electrostatics, polymer structures, elasticity, and heat transmission. Add to this process chefs like yours truly, and the results could be inspiring.

Aside from gaining satisfaction and pride from my initial prediction, I'm pleased to see that the pursued objectives have been achieved. The scientific concepts mentioned above are now gathered almost entirely in the pages of this book, and the chef recipes add an inspiring, experimental component that helps explain the science. What's more, the different chefs who have been participating in the Harvard course over these ten years have found a space in which to pose inquiries and bring up some of the challenges encountered in the kitchen. The lab gives us a way to express new ideas that we would like to develop, all of which requires scientific knowledge and research.

For our part, these scientific questions had been with us in the chef community for some time. The incorporation of science into elBulli's creative system came about explicitly when, at the beginning of this century, two scientists, Pere Castells and Ingrid Farré, joined the kitchen team as direct collaborators. It was what we called elBullitaller's Scientific Department, which yielded very interesting results, although surely more from a culinary than a scientific perspective. After some time, this attitude of scientific interest led to the creation of the Alícia Foundation, which in a way represented the desire to professionalize this dialogue between cuisine and science. Then, Harvard categorically opened the doors to an even more open dialogue when they approached us about the course. This has been a bidirectional conversation from the beginning, which I believe benefits both parties. On the one hand, gastronomy has facilitated the understanding, or at least the explanation, of the physicochemical reactions that take place in the kitchen. On the other hand, science has found, or so I like to believe, a new way to convey a diverse set of concepts, from the perspective of its more practical application. With this symbiosis, gastronomy gained access to knowledge, both to understand and to develop new sensory experiences in the tasting, while at the same time it became a promising new creative tool for science education.

Combining science with cooking, and learning from their respective viewpoints, establishes a very fruitful dialogue. Although we have been cooking and searching for universal knowledge for millennia, these exercises have intersected only sporadically in human history. The turning point likely came in the last quarter of the twentieth century, when a group of scientists started experimenting with what they called molecular gastronomy: a scientific attempt to explain why different reactions take place to make a series of culinary preparations possible.

Importantly, understanding the composition of the products and the reactions produced by culinary techniques has been beneficial beyond improving haute cuisine and dining experiences. It has also made it possible to better determine why certain allergies and intolerances occur—this is a phenomenon of growing importance for society in general, and catering in particular, often to the extent that certain tables will require different menus. Furthermore, advanced knowledge of products and preparations not only allows us greater innovative and creative ability, but being aware of everything that can happen on the microbiological scale also contributes to achieving increased food safety and hygiene.

All of these aspects, and many more, have been addressed for the last ten years in the Harvard course, and the majority of them are included in this book. The authors aim to contribute to the general public's understanding of the connections that have historically taken place between science and cooking, and the current state of these collaborations and dialogues. However, if we consider the future, there is still much to be done. Why, for example, does Harvard—along with other prestigious universities—have a school of architecture, but not one of gastronomy? This is due to a cultural bias that needs to be corrected, and I am convinced it will be rectified in the near future, since we know gastronomy has significant intellectual potential.

At the elBullifoundation, one of the most important projects that we have been promoting for years is the deepening of this dialogue and the development of a global project to maximize opportunities and strengthen the synergy between cooking and science, always with the aim of deepening our culinary knowledge and professionalizing our discipline to the highest level. I am convinced that this commitment to knowledge has resulted in the most prepared and brilliant generation of future chefs and restaurateurs in history. I have no doubt that Harvard's continued work in pursuit of this dialogue will enable us in the future to have chefs capable not only of understanding their trade and the history of their discipline, but also of putting into practice the links established between gastronomy and science.

Ferran Adrià

Foreword

When Ferran asked me to join him at Harvard's Science and Cooking course almost a decade ago, we knew it was something special. Every chef who has worked with Ferran, and many others around the world, know this truth: that cooking is science and that science is cooking. Many amazing discoveries have been made in history by scientists working with food, as well as by chefs employing the scientific method. It was Nicolas Appert, a French pastry chef and chemist, who invented the process of canning in the early nineteenth century through a series of experimentations, leading to an entirely new way to safely preserve and consume food. Microbiologist Louis Pasteur's pasteurization process revolutionized the food industry—and has probably saved millions of lives since its discovery. Recent discoveries about fermentation and preservation by places like Noma's Fermentation Lab are leading a new revolution into the science of microbes, and may be key to figuring out how we will be feeding the next billion people on Earth.

We are thinking about scientific processes every day at my restaurant minibar, where we use physics, chemistry, microbiology, fluid dynamics, and more to create dishes that are new and interesting—and, of course, delicious. But this type of experimentation isn't just happening at minibar. Scientific thought enters into every one of my restaurants—from how long we boil vegetables (and at what concentration of salt) at Beefsteak to the type of wood we use for smoke at America Eats Tavern to the heat diffusion in our woks at China Chilcano. When we use the term "molecular gastronomy," we are really describing all cooking, because all of it involves the manipulation of molecules. For example, boiling, steaming, and making ice are all transformations of water molecules. Thinking about science as cooking and vice versa is crucial for everyone from the home cook to the top chef, from your stovetop to the kitchen at elBulli.

Ferran is right when he says that every prestigious university in the world should have a gastronomy school—it is important for students to have the opportunity to understand how food connects to every single part of our lives. Here we are looking at food as science, but food is also history, culture, diplomacy, national security, and much, much more. And it would make our society stronger if we all were better aware of where our food comes from and what deep impact it has on the world around us. It's why this book—and the Harvard course on which it's based—is such a valuable resource. Engaging for anyone and everyone, not just students of science, *Science and Cooking* can help readers begin making connections and building a systemic view of the fascinating world of food. This is vital for our future: the next revolution in science could very well come from a young culinary student, and on the other hand, the world of cooking may be rocked by a new discovery in quantum dynamics or astrophysics. We must learn from one another and keep pushing these conversations forward—and to me, this book is the perfect way to do so.

José Andrés

Introduction

As professors who teach physics and chemistry, we have long been motivated by the dream that anyone would want to learn science if it were placed in a compelling context. For this, what could be a better topic than food and cooking? Every day we eat; we buy ingredients and cook them through detailed protocols called recipes, and if we do everything correctly, we are rewarded with a delicious meal. Social media is full of picturesque dishes shared for others to admire. But how do these recipes really work? Cooking has long been believed to be an empirical subject—"just follow the instructions!" But in fact, there's a reason why the recipes that we use work the way that they do—the answers are rooted in the fundamentals of physics and chemistry. We believed that by creating an environment where our students could figure out for themselves why recipes worked, they would both be empowered with skills to improve their own cooking and also gain an appreciation for the scientific method—all the while, learning some physics and chemistry. Together, we have been teaching a class called "Science and Cooking: From Haute Cuisine to Soft Matter Physics" at Harvard College for the last decade, and online through HarvardX since 2013. As more and more students enrolled both in person and online, we felt the need to share these insights with a wider audience. For this reason, we have written this book: to teach you remarkable science through simple, engaging experiments on food and cooking.

Think about it: Why do we cook chocolate chip cookies for 10 minutes and not 20 minutes? Why does steak taste differently cooked at different temperatures? Why do we knead bread? What determines the amount of egg required for making mayonnaise? Although these recipes might have been discovered empirically, the reasons for these rules are strongly rooted in science and the scientific method. We created the class together with our colleagues and friends Ferran Adrià and José Andrés, two of the most inventive chefs on the planet, both of whom dreamed that the discoveries and innovations they had used so successfully in their restaurants could inspire people to learn to think about science.

Ferran revolutionized cooking and haute cuisine through his legendary restaurant elBulli, which was widely heralded as the best restaurant in the world. He has likely invented more completely original recipes and techniques than anyone who has ever lived—ranging from hot ice cream to culinary foams to spherification and beyond. He did this by figuring out how to repurpose natural ingredients with other uses to invent an incredible array of new foods. These recipes literally deconstructed the essential elements of older recipes and rebuilt them from scratch. From the point of view of a scientist, Ferran's reconstructions are beautiful demonstrations of the scientific method. He figures out how recipes work and then uses only the essential ingredients in the right proportions to make new dishes, even more spectacular than the originals. In reinventing these dishes, he at the same time shows exactly why the recipes worked in the first place. Ferran and José knew that the culinary innovations of the last decade were consequential, and they wanted these innovations to be featured and used in the classroom. Thus, they proposed that our Science and Cooking class should prominently feature the creations of world-famous chefs who could show their magic to our students and help us scientifically deconstruct recipes while learning science. This was a bold idea: leading universities have long supported creativity, through artists, authors, and architects. But until this class, we are unaware of an instance where novel creations in cooking were pushed to the forefront.

The first year's roster was headed by the very best chefs in the world, including José Andrés, Wylie Dufresne, Joan Roca, Grant Achatz, Dan Barber, Carme Ruscalleda, Nandu Jubany, Joanne Chang, Carles Tejedor, Enric Rovira, Bill Yosses, and David Chang. Each week had multiple sessions: On Monday night, the visiting chef would give a free lecture to the general public in Cambridge. Lines formed for these lectures hours before doors opened, and chefs often brought an entourage to prepare culinary samples for the audience. Lectures were scheduled for an hour, but would often last for hours beyond this, such was the curiosity of the general public. On Tuesday, the visiting chef would give a version of this lecture to the Harvard class, illustrating the scientific theme of the week by cooking spectacular recipes. Samples were often served. On Thursday we would delve in depth into the scientific concepts. Each week there was also a laboratory session, in which the students would cook recipes illustrating the scientific theme of the week. They would work on problem sets, make scientific measurements on these recipes in the lab, and end by eating the result of their creation.

A major inspiration for the class was Harold McGee, the famed author of the classic book *On Food and Cooking*. Published in 1984, it remains the best book to explain the scientific basis of cooking. Our own copies are worn around the edges. Harold's work is universally admired by chefs around the world, whose copies are similarly worn. We were very fortunate that Harold agreed to participate in the class from the very beginning, and has served as a mentor, a sounding board, and a source of wisdom to us ever since.

Since its inception, thousands of students have taken the Science and Cooking class at Harvard, and hundreds of thousands of students around the world have taken the online class—in Brazil, China, Britain, India, Saudi Arabia, Japan, and more. Like the class, the book is organized around scientific themes. Each chapter explains the basic principles, motivated by both common recipes and those of today's leading chefs. There are experiments that you can do at home, but even if you don't do them all, our hope is that you'll join us in recognizing the remarkable science behind food and cooking. Ready? Let's get to it!

Science
and Cooking

CHAPTER 1

What Makes a Recipe?

Science isn't just a matter of specialized knowledge about molecules or math, it's a way of thinking. Above all, it's a way of thinking. It's about observing carefully, being curious about what you see, noticing an interesting or unusual or unexpected phenomenon, and then finding ways to understand that phenomenon through controlled experiments—coming up with comparisons of different ways of doing things that teaches you something about what's going on.

—Harold McGee

Recipes are instructions for cooking, and you are meant to follow them carefully. They give you the specific amounts of ingredients and the steps you need to take to make your dish, transforming the raw ingredients into something delicious. The process, as you'll see, can be quite magical. Consider, for example, the sidebar with chocolate chip cookie recipes.

Standard Chocolate Chip Cookies

Ingredients

270 g (2¼ cups) all-purpose flour, sifted

1 teaspoon baking soda

1 teaspoon salt

225 g (1 cup, 2 sticks) unsalted butter, at room temperature

150 g (¾ cup) granulated sugar

150 g (¾ cup) packed brown sugar

1 teaspoon vanilla extract

2 large eggs

340 g (2 cups) chocolate chunks, chips, or wafers

125 g (1 cup) chopped nuts

Directions

1. Preheat the oven to 375°F (191°C).

2. Whisk together the flour, baking soda, and salt in a small bowl.

3. Beat together the butter, sugars, and vanilla in a large bowl until lightened and creamy. Add and beat in the eggs, one at a time. Gradually beat in the flour mixture. Stir in the chocolate and nuts.

4. Drop ping pong–size balls of dough onto ungreased baking sheets, using an ice cream scoop or rounded tablespoon.

5. Bake for 9 to 11 minutes, until golden brown. Cool the cookies on the baking sheets for 2 to 3 minutes, then transfer to wire racks to finish cooling.

Christina Tosi's Cornflake Chocolate Chip Marshmallow Cookies

Ingredients

225 g (2 sticks) unsalted butter, at room temperature

250 g (1¼ cups) granulated sugar

150 g (⅔ cup tightly packed) light brown sugar

1 large egg

2 g (½ teaspoon) vanilla extract

240 g (1½ cups) all-purpose flour

2 g (½ teaspoon) baking powder

1.5 g (½ teaspoon) baking soda

5 g (1½ teaspoons) kosher salt

270 g (3 cups) Cornflake Crunch (recipe follows)

125 g (⅔ cup) mini chocolate chips

65 g (1¼ cups) mini marshmallows

Directions

1. Combine the butter and sugars in the bowl of a stand mixer fitted with the paddle attachment and cream together on medium-high for 2 to 3 minutes. Scrape down the sides of the bowl, add the egg and vanilla, and beat for 7 to 8 minutes.

2. Reduce the mixer speed to low and add the flour, baking powder, baking soda, and salt. Mix just until the dough comes together, no longer than 1 minute. (Do not walk away from the machine during this step, or you will risk overmixing the dough.) Scrape down the sides of the bowl with a spatula.

3. Still on low speed, paddle in the cornflake crunch and mini chocolate chips just until they're incorporated, no more than 30 to 45 seconds. Paddle in the mini marshmallows just until incorporated.

4. Using a 2-ounce ice cream scoop (or a ⅓-cup measure), portion out the dough onto a parchment-lined sheet pan. Pat the tops of the cookie dough domes flat. Wrap the sheet pan tightly in plastic wrap and refrigerate for at least 1 hour, or up to 1 week. Do *not* bake your cookies from room temperature—they will not hold their shape.

5. Preheat the oven to 375°F (191°C).

6. Arrange the chilled dough a minimum of 4 inches apart on parchment- or Silpat-lined sheet pans. Bake for 18 minutes. The cookies will puff, crackle, and spread. At the 18-minute mark, the cookies should be browned on the edges and just beginning to brown toward the center. Leave them in the oven for an additional minute or so if they still seem pale and doughy on the surface.

7. Cool the cookies completely on the sheet pans before transferring to a plate or to an airtight container for storage. At room temperature, the cookies will keep fresh for 5 days; in the freezer, they will keep for 1 month.

Cornflake Crunch

Ingredients
170 g (½ of a 12-ounce box) cornflakes
40 g (½ cup) milk powder
40 g (3 tablespoons) sugar
4 g (1 teaspoon) kosher salt
130 g (9 tablespoons) unsalted butter, melted

Directions
1. Preheat the oven to 275°F (135°C).

2. Pour the cornflakes into a medium bowl and crush them with your hands to one-quarter of their original size. Add the milk powder, sugar, and salt and toss to mix. Add the butter and toss to coat. As you toss, the butter will act as glue, binding the dry ingredients to the cereal and creating small clusters.

3. Spread the clusters on a parchment- or Silpat-lined sheet pan and bake for 20 minutes, at which point they should look toasted, smell buttery, and crunch gently when cooled slightly and chewed.

4. Cool the cornflake crunch completely before storing or using in a recipe. Stored in an airtight container at room temperature, the crunch will keep fresh for 1 week; in the fridge or freezer, it will keep for 1 month. ✽

We are so used to making and eating cookies that we don't usually recognize the wondrous magic of these recipes. Think about it: we transform a set of simple, boring ingredients into a delicious morsel that looks nothing like the ingredients from whence it came. The texture, color, and especially taste are all different. What starts off as a mushy dollop on your baking sheet becomes a crisp, sometimes crumbly, sometimes soft (depending on your recipe) bite that melts in your mouth.

Much of cooking works this way. The first cookie recipe was invented about two hundred years ago—a plain, dry sugar cookie recipe with minimal fat. Over the centuries since, people have improved on the recipe—the most conspicuous improvement being the introduction of more and more fat. Today, the internet abounds with cookie recipes. (You have probably made many of them yourself.) The core principle, however, has remained nearly the same: Combine 2 parts flour, 2 parts fat, and 1 part sugar and mix well. Shape the batter into balls, bake at the right temperature for the right length of time, and voilà, you get cookies. Changing the percentages of the major ingredients means you might create an entirely different substance, sometimes even a new recipe. For example, if you use twice the amount of sugar as flour, you get brownies. When this was discovered, at the dawn of the twentieth century, it must have been a surprising and marvelous advance—bakers achieved a new texture completely different from that of cookies. Cooking isn't an entirely static enterprise: even though cookie-making is a very old subject, sometimes modern technology makes advances possible. Our favorite example is when Christina Tosi, pastry chef and founder of Milk Bar, recently invented a novel variant on the cookie recipe in which she figured out how to use much more fat than ever before possible, with a clever use of an electric mixer; the recipe is shown in the sidebar. We will discuss her work more thoroughly in chapter 6.

If you skim through this book, you will notice that the pages contain many recipes. While it's not required to make anything yourself to learn the science, we nonetheless encourage you to stop and make any recipe that appeals to you. This is what we often do in our Science and Cooking class at Harvard—as recipes come up, we have lab sessions in which the students stop and cook. For example, you could stop reading right now and go make the chocolate chip cookies. You can make the normal ones, then make Christina's version, and compare them. Which cookie recipe tastes better, and why? Indeed, eating is an excellent way of measuring the differences in the recipes yourself. You can also try out new variants of the recipe to compare and contrast. This is tasty and fun, and it's the very essence of cooking. Our recipes are for common foods that you might make every day, as well as more adventurous dishes from the most creative chefs of our time.

If, however, you find yourself constantly cooking while reading this book, please remember that this is ultimately not a cookbook. Yes, we want you to experiment with recipes. But the purpose of experimenting is to convince you that by asking the right questions, it is possible to *understand how recipes work and why the instructions and amounts are chosen as they are.* Recipes are not arbitrary. In fact, the instructions and amounts are strongly rooted in scientific principles. While many recipes were discovered by accident, they also come with years and generations of tweaks. Like much of science, there is much about even common recipes that we do not understand. Nicholas Kurti famously stated, "I think it is a sad reflection on our civilization that while we can and do measure the temperature in the atmosphere of Venus, we do not know what goes on inside our soufflés."

Despite the rise in scientific approaches to cooking, the situation is not much better today.

But if we proceed with curiosity and humility, and ask the right questions, we can unravel the mystery that a recipe holds and understand why it is designed in the way that it is. Sometimes it will turn out that the scientific underpinning of a recipe is easy to figure out; other times it will turn out to be more complicated. But in either case, we will show you that by asking the right questions and using bits of science, it is often possible to get to the heart of a recipe. This will make you a better chef, enabling you to recognize themes that come up again and again throughout cooking. It will make you understand how dishes that appear completely different (for example, al dente pasta and a medium-rare steak) can have deep scientific similarities. It will also enable your own creativity in the kitchen, by making you think about how subtle changes can make a big difference.

The underlying goal of this book goes far beyond cooking. Being curious, and learning to ask the right questions to deconstruct a complex process, is at the heart of science and the scientific method. A large part of being a scientist is having the courage to ask hard questions, the humility to admit when you are wrong, and the doggedness to find the answers, by hook or by crook. And as in all things in life, when you don't succeed, try again. Even the simple and classic problem of the chocolate chip cookie has proven remarkably difficult to crack. Nonetheless, this is an attitude that will improve your own cooking—and also give you a window into how scientists actually work. It is important to emphasize that you do not need to be a professional scientist to have the curiosity to ask why, and to use the resources at hand to ask questions and find answers.

Deconstructing Recipes

How do we go about deconstructing a recipe? Let's consider the case of the chocolate chip cookie. Every recipe has two basic parts: the ingredients and the cooking process, known as the method. The crux of understanding a recipe is to determine how the process transforms the ingredients into a substance with a completely different character. For chocolate chip cookies, we start with a bunch of powders (flour, sugar, salt), some liquid (in the form of eggs), and some fat in the form of butter. Then by combining them according to the right process, we arrive at cookie dough—a substance whose properties differ from the ones we started out with. You can stretch cookie dough, make it into balls and squish them, and even play catch with it. Try to do the same thing with any of the initial ingredients. It just won't work. The dough also tastes much better than the raw ingredients. When you put it in the oven, it transforms again, this time into a foamy solid with a satisfying texture. These transformations result from the choices that we make regarding the process and the ingredients.

To deconstruct a recipe, we must first understand what the ingredients are made of. For most foods, you could find this out quite easily by simply looking at the nutrition label, showing how much fat, protein, and carbohydrates are in the food. This is important because from a caloric standpoint, fat contributes 9 Calories/gram, while protein and carbohydrates contribute 4 Calories/gram. Thus, if you are calorie counting, it is best to limit your fat intake.

But this crude characterization misses the most interesting point. Fats, proteins, and carbohydrates are *molecules*, with shapes, sizes, and properties that are quite different from each other. A key question in understanding a recipe is to figure out the molecular transformations of the ingredient molecules, why the transformations occur, and how they affect the final product. In this book, we have a mantra that we will follow again and again:

Understanding a recipe requires understanding how the ingredient molecules are transformed into the molecular structure of the final recipe.

Arguably, the most important characteristic of a food is the way we perceive it when we eat it. Optimizing our experience with food is the reason we spend so much effort cooking it. There are two aspects of our sensory experience: texture and flavor. Imagine eating a wet cookie—the flavor might be right, but you won't like it. At the same time, imagine eating a burned cookie. The texture could be perfect, but you will spit it out. A fascinating fact is that the molecular properties that lead to texture and flavor are entirely different and largely come (with a few important exceptions) from different types of molecules. In this book we will refer to these different molecule types as "texture molecules" and "flavor molecules." We will see when we examine recipes that they behave quite differently; indeed, the requirements for producing outstanding flavor *and* texture are part of what makes good cooking so difficult.

Before we proceed, and since we're on the topic of molecules, a word about "molecular gastronomy." This has been a trendy but sometimes pejorative term that has been levied at the most creative chefs of our time. These chefs indeed work with molecules, and they became famous for discovering entirely new natural ingredients and processes for changing flavor and texture. We will discuss some of their discoveries in this book. We want to emphasize, however, that the use of molecules in cooking is not new: *all cooking is molecular.* As our friend José Andrés famously said, "Grating parmesan cheese is a molecular exercise." To label the creative cooking of today "molecular" without understanding that all cooking is molecular is to completely misunderstand what cooking is all about.

Texture Molecules

Let's return to our two types of molecules. The texture molecules are those found on nutrition labels: proteins, fats, and carbohydrates, as shown in Figure 1. The remarkable transformations in food that occur as we cook are almost entirely due to these molecules, and they are dramatically different for carbohydrates, fats, and proteins. We will see evidence of this throughout this book. To whet your appetite with some examples, consider the fact that fats don't dissolve in water— oil and water famously don't mix. (Think of a vinaigrette that has been left to stand: the oil separates on the top and the vinegar on the bottom.) By contrast, carbohydrates like sugar dissolve very easily: this might be hard to believe, but at room temperature, a cup of water can dissolve twice its weight in sugar! Candy making, as we'll see later on, is all about controlling the sugar-water ratio. It works because you can pack even more sugar into the water when you heat the mixture. At 100°C (212°F), water holds *four* times its weight in sugar. When you then mix fats into the sugar water, it transforms the textures and tastes and can result in

Fat

Carbohydrate

Protein

Nutrition Facts

Serving Size: 1/4 cup (31 g)

Amount Per Serving

Calories 110	Calories from Fat 0

	% Daily Value*
Total Fat 0 g	0%
Saturated Fat 0 g	0%
Trans Fat 0 g	
Cholesterol 0 mg	0%
Sodium 0 mg	0%
Potassium	
Total Carbohydrate 23 g	8%
Dietary Fiber	
Sugars	
Sugar Alcohols 0 g	
Protein 3 g	
Vitamin A 0 IU	0%
Vitamin C 0 mg	0%
Calcium 0 mg	0%
Iron 1.08 mg	6%

FIGURE 1 Nutrition labels are designed to communicate the nutritional content of a food. The nutritional content is directly related to the molecular composition, so we can learn quite a bit about the content of various texture molecules by looking at the reported values. In fact, sometimes it is possible to "reverse engineer" a recipe's ingredients using a combination of the nutrition facts and the ingredients, which are listed in descending order by weight. The pictured nutrition label is an example of a label in the United States. Labels in other countries are similar in content, although they may be written in different styles and using other units.

The major texture molecules are indicated by the arrows. Fats at the top, followed by carbohydrates about halfway down, and then protein. Next to each one is listed how many grams of those molecules can be found in each serving. Fats, carbohydrates, and proteins are all relatively large molecules, and being texture molecules, there needs to be enough of them to affect the texture. Besides water, they constitute a majority of the weight of our food. In this example, carbohydrates and proteins together make up 26 g in a 31 g serving, which means about 5 g is water.

delicious caramels. Proteins, however, are completely different. They dissolve in water, but their cooking superpower is that when you heat them, they fall apart and then stick back together again—leading to a total transformation.

The final molecule that contributes to texture is the dominant ingredient: water. (Ironically, it also happens to be commonly left off nutrition labels.) You may not think so when looking at a steak or a potato, but more than half of it is water— 60% for the meat and 80% for the potato. Even flour, one of the seemingly driest foods, contains as much as 15% water. See Figure 2 for some other examples. As it turns out, when we change the texture of foods, we are often primarily manipulating water content. This makes the properties of water critically important. Since food is mainly water, the laws that govern heating a steak are exactly the same as those for cooking a potato or baking a cake. These laws are essentially the same as for heating a cup of water.

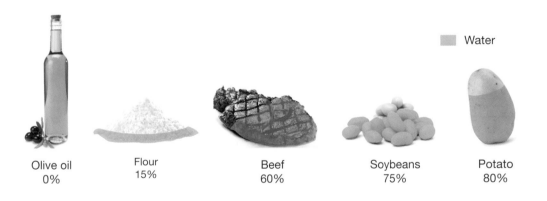

Water

| Olive oil | Flour | Beef | Soybeans | Potato |
| 0% | 15% | 60% | 75% | 80% |

FIGURE 2 Unsurprisingly, oil contains no water, but most foods that come directly from living organisms are mostly water. Flour looks dry, but starches always absorb a certain amount of water from the atmosphere, and it is extremely difficult to keep them fully dehydrated.

Flavor Molecules

Despite their importance in our experience of food, flavor molecules are not listed on nutrition labels. One important reason is that flavor molecules are tiny and appear in minuscule amounts—sometimes there are as few as a millionth or a billionth as many flavor molecules as there are proteins, carbohydrates, and fats. But without them food would be boring and, well, flavorless.

There is an almost infinite number of flavor molecules, and the flavor of any one food can be a result of the combination of hundreds of these molecules. This diverse array is simplified somewhat by the fact that all flavor molecules can be categorized into two broad classes (see Figure 3), which differ both by how we sense them and also by how we produce them for cooking. *Taste molecules* bind to the taste receptors on our tongues. There are five in total: sweet, salty, sour, bitter, and umami. *Aroma molecules*, on the other hand, bind to the olfactory receptors at the back of our noses. They get to the receptors by detaching from the food when we chew it and floating in the air through the back of our mouths to the back of our noses. Aroma molecules are a major source of the rich and varied flavors of foods. Very few animals can sense them; it has even been argued that the evolution of the human brain is a direct response to our ability to detect them.

In general, both taste and aroma molecules tend to be very small. The aroma molecules in particular have to be light enough to float in the air; we say that they are volatile. But even the taste molecules are generally small, since small molecules tend to bind better to the taste receptors. By contrast, proteins, fats, and carbohydrates are large and cumbersome, and as a result generally have very little flavor on their own—they are too large to bind to the receptors. Instead, they proceed directly to our stomachs, where they are incinerated for our caloric intake.

It can be surprisingly difficult, however, to tell the difference between taste and aroma. Together, they constitute flavor. But the aroma molecules of foods easily slip up through the back of our mouths into the back of our noses, giving us the impression that something tastes a certain way, while really it is the aroma. For an exercise in distinguishing between the two, try the first experiment in the sidebar. The second sidebar experiment shows that the combination of tastes is quite complicated and interesting. Both of these exercises highlight facts about flavor that are important for understanding how to make tasty food.

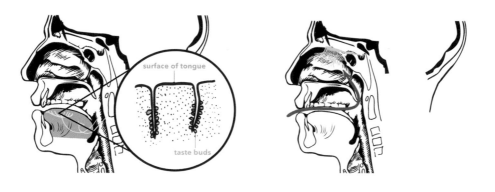

Taste molecules

Sweet
(potentially high energy;
sucrose is detected at 10 mmol/L)

Sour
(potentially either bad or fresh food)

Salty
(electrolyte)

Bitter
(potentially toxic;
quinine is detected at 8 µmol/L)

Umami
(savory, potentially high energy food)

Aroma molecules

sulfurous · green
celery · alliaceous · grassy · fruity
soupy · ester-like
mushroom · citrus
earthy · terpenic
dairy · minty
buttery · flavor matrix · camphoraceous
fatty · floral
rancid · sweet
meaty · spicy
animalic · herbaceous
bouillon · woody
caramel · roasty · smoky
nutty · burnt

16 1 2 3 4 5 6 7 8 9 10 11 12 13 14 15

surface of tongue

taste buds

FIGURE 3 Taste molecules, shown on the left, are sensed by receptors on the tongue. Although there are five main tastes, our experience of them varies depending on how they are combined in different foods. We have evolved to associate certain tastes with good or bad effects on our bodies. Sweet, which is a pleasurable sensation for most people, tends to be associated with fast, high energy food sources. Bitterness, by contrast, can help identify something toxic. Perhaps this difference is why it takes almost no effort to coax a child to finish a sweet dessert, yet bitter foods often fall into the category of "acquired tastes."

Aroma molecules, shown on the right, are sensed by the olfactory receptors in our noses. They can get there via two routes: through our nostrils and, perhaps even more importantly, through the back of our mouths as we chew and swallow the food. There are some eight hundred genes for olfactory receptors in humans, which makes it a much more complex and sensitive system than taste. In fact, the perceived differences between many foods can be traced to their aroma molecules, not their taste molecules. A famous experiment has blindfolded people hold their noses while eating pieces of apple, onion, and potato. The tasters are told to guess which of the three foods they are eating. What tends to happen is that the foods taste very similar until they are about to be swallowed. Then, when the aroma receptors sense the food passing through, you get the final identity confirmation that the blocked nose missed. If a taster has a cold, the receptors in both the nose and the mouth are blocked by mucus, which causes all food to taste bland.

SIDEBAR 2: PEPPERMINT CANDY

Hold your nose and use the back of your tongue to close off the back of your throat as much as you can. Then, put a peppermint candy in your mouth and try to identify the flavor(s) that you taste. Next, let go of your nose—what happened? When you held your nose, the candy probably just tasted sweet. Then when you let go, there was a blast of minty flavors. This is because menthol (the flavor molecule in mint) consists of the minty smell we are all familiar with, plus a hint of bitterness and a cool sensation. When you held your nose, your body was unable to detect the menthol aroma, and the bitterness was likely overwhelmed by the sugar in the candy. The cooling sensation comes from the menthol triggering certain nerves in the nose and mouth. By blocking your nose, you are also preventing this from happening to its full extent. You can try this exercise with any food to separate the smell of a food from its taste, which will help you appreciate how they all work together to create unique flavors. ⚛

SIDEBAR 3: BALANCING FLAVOR: SUGAR-ACID

Soft drinks are known for having a lot of sugar in them. In fact, in each 12-ounce/355 mL can—about 1½ cups of soda—there's about ¼ cup of sugar. But, have you ever considered how it's even possible for manufacturers to add such high amounts of sugar without the drink tasting too sweet? To answer this question, let's perform a quick experiment.

Fill a glass with drinking water and stir in 1 teaspoon of sugar at a time, taking a sip after each addition, until the sugar water becomes too sweet to be enjoyable. Next, stir in ¼ teaspoon of vinegar at a time, again taking a sip after each addition, until the drink tastes drinkable to you. Be sure to keep track of the amount of vinegar you added to the sugar water.

Fill a separate glass with drinking water and add the same amount of vinegar you added to the previous glass and take a sip—and try not to make a face!

This is the secret of Coca-Cola. It contains far too much sugar for most people to find tasty. However, by adding acid (and other flavors), the flavor components combine to produce a pretty delicious drink. Carbonation is another source of acidity, which is why a flat Coke tastes sweeter than a fresh one.

This is just one example of the myriad ways in which taste molecules can balance each other. Many recipes exploit this fact to add layers of flavor and bring out hidden tastes from the ingredients. The crafty cook is aware of how different flavors play off each other and is not afraid of experimenting with them to improve the overall flavor of their food. For example, the sugar and acid effect is why some cooks add a bit of sugar to their marinara sauce to balance tomatoes that are too acidic. Nathan Myhrvold even goes so far as to add a pinch of salt to red wine to make it taste better. ❀

WHERE DO FLAVOR MOLECULES COME FROM?

Flavor molecules occur naturally in food, but they are also added by the process of cooking. You might be surprised to learn that cooked food contains many more flavor molecules than that same food uncooked. But consider: Baked cookies taste different than raw cookie dough. The taste of a cooked steak is unlike that of steak tartare. Recall that protein, carbohydrate, and fat molecules are far too large to bind to our taste and aroma receptors, so we must use tiny molecules to add taste to food. Sometimes we add these in the early stages of a recipe. For example, in a chocolate chip cookie recipe, we add sugar, salt, and vanilla. Salt is tiny enough to bind to the salt receptors on our tongues, and vanilla contains a small molecule called vanillin that binds to the odor receptors at the back of our noses. Chocolate also has flavor, but it is a more complex fermented food, with properties we will consider later on.

But the real magic of flavor creation happens when we cook the food. The process of cooking can literally break apart the protein, carbohydrate, and fat molecules and turn them into flavor molecules! Those large molecules break into small ones, which are then broken into even smaller ones and then even smaller ones. Eventually, they are small enough that they can be detected by our taste and aroma receptors. One key agent in this flavor creation is heat, which causes the molecular breakdown (we'll discuss heat in more detail in chapter 2). But similar flavor creation happens in many other cooking processes, too. Cooking food with microbes, as with various food fermentations like sauerkraut or pickles, also breaks down large molecules into tiny flavor molecules. The same is true for smoking or ageing.

FLAVOR AND ACIDS

The simplest example of a common flavor molecule is acid, which is responsible for sour tastes. The defining property of acidic molecules from a scientific standpoint is that they easily shed hydrogen ions. Our taste buds sense acids by detecting the hydrogens: the hydrogen ions block the proton channels in the taste buds and send a "sour" signal to the brain. Remarkably, it takes very few hydrogen ions for a food to be detected as sour. Consider lemonade: Its main acid is citric acid, consisting of six carbons, seven oxygens, and eight hydrogens. When citric acid dissolves in water, it breaks into pieces, leaving small amounts of hydrogen ions floating around. Indeed, if you dissolve even a very small amount of citric acid in water, there will be about a million times fewer hydrogen ions than water molecules. It would seem that the hydrogen ions would rarely bump into a taste bud, but in fact they make constant contact with taste buds and are all that's needed to trigger a sour taste. This truly remarkable scientific phenomenon is a general property of flavor: it takes comparatively few flavor molecules to produce a strong taste sensation. Thus cooking well requires us to very carefully balance flavors.

Acids are important throughout all of chemistry, not just when it comes to flavor. In fact they are so important that scientists have developed a special scale, called the pH scale, to measure how strong an acid is. The pH measures the number of hydrogen ions as a percentage of the total number of water molecules. For historical reasons, scientists measure this using the scale explained in the sidebar. The important point is this: Pure water has a pH of 7, whereas lemon juice has a pH of about 2. This means that there are five orders of magnitude (100,000) more hydrogen ions in lemon juice than there are in pure water. In other words, whereas in lemon juice there are a million times fewer hydrogen ions than there are molecules of water, in water there are 100 billion times fewer hydrogen ions than molecules of water.

pH values for some common liquid foods
Below 7 = acid, 7 = neutral, above 7 = basic

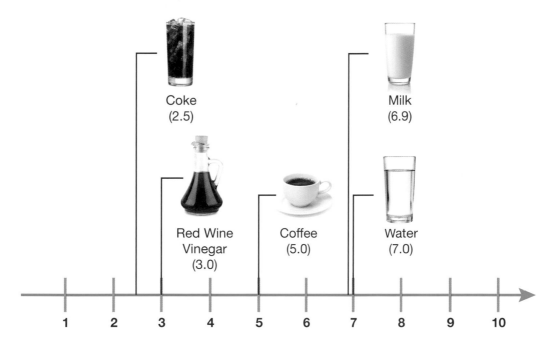

pH is a logarithmic scale (or simply "log scale") that measures acidity (and alkalinity). It tells you how many protons (H^+ ions) are in a solution. Why is this important to us? H^+ ions essentially clog up, or stop, proton channels in our taste buds, thereby sending the signal for "sour" to our brains. Solutions that have low pH (pH < 7) are acids and taste sour to us, while alkaline solutions (pH > 7) often taste bitter, metallic, or soapy.

The pH scale, developed by a Danish scientist named Søren Sørensen, is based on concentration of hydrogen ions, which is given as the number of moles of H^+ per liter of solution; 1 mole is about 6×10^{23} units (ions, in this case)—that's 6 followed by 23 zeros! We use this strange number—it even has a name: Avogadro's number—because it makes the math easier to handle. Otherwise, counting the number of ions or molecules quickly gets very cumbersome. The logarithmic scale tells us that every unit change in pH means that the H^+ concentration has changed tenfold. Lemon juice, with a pH of 2, has ¹⁄₁₀ the number of H^+ ions as our gastric fluid, which has a pH of 1. Together, this means that we can determine the pH by looking at the exponent of the concentration. For example, a liter of water has 10^{-7} moles of H^+ ions at room temperature, so a neutral pH is defined as 7. (It is worth noting that Sørensen discovered his scale when studying the fermentation of beer—a worthwhile topic for Science and Cooking!)

Early on, though, people realized that total hydrogen concentration does not always directly translate into acidity; instead, the dissolved or active H^+ concentration is what's important. ⚛

SIDEBAR 5: DANIEL HUMM'S DUCK SAUCE

Daniel Humm, the remarkable chef of Eleven Madison Park in New York City, has for years been rated one of the top chefs in the world. He is a great aficionado of acids in cooking and uses them to great effect in his restaurants. He explained to our Harvard class that acids have the ability to make even humdrum ingredients taste special—and to make top-quality ingredients truly extraordinary. For example, this is the secret behind salt and vinegar potato chips: the acid from the vinegar is what makes you keep coming back for more. It is also the secret behind fast food, cheap wine (just add acid!), soda . . . the list goes on.

To illustrate Daniel's dictum that acids can make delicious food taste even better, let's examine his recipe for Citrus Duck Jus (sauce for his famous Lavender and Honey Duck), which contains *four* different acids: lemon juice, lime juice, orange juice, and vinegar. In our class, we assign a homework problem for the students to calculate the pH of the resulting sauce. Daniel has determined that the perfect pH for his sauce is 4.6. The pH of this sauce is particularly important because it balances the gamy, fatty, rich flavor of the duck and the sweet flavor from the sugar in the sauce.

Citrus Duck Jus

Ingredients

50 g canola oil

300 g duck necks, wings, and feet

1 kg Chicken Jus (recipe follows)

50 g Citrus Gastrique (recipe follows)

10 g lime juice

5 g orange juice

3 g raspberry vinegar

7 g salt

Directions

1. Heat the oil in a large saucepan over high heat. Add the duck necks, wings, and feet and sear them, turning occasionally, until thoroughly caramelized, about 20 minutes.

2. Drain off any rendered fat from the pan and add the chicken jus. Bring to a simmer and cook until reduced to sauce consistency. Add the gastrique and stir to combine.

3. Strain the sauce through a chinois and add the lime juice, orange juice, raspberry vinegar, and salt. Keep warm.

Chicken Jus

Ingredients

100 g canola oil

560 g diced onions (2 cm/¾-inch pieces)

260 g peeled and diced carrots (2 cm/¾-inch pieces)

260 g diced celery (2 cm/¾-inch pieces)

100 g tomato paste

1 (750 mL) bottle dry red wine

4.5 kg chicken wings

2.5 kg chicken feet

13.5 kg water

2 bay leaves

10 thyme sprigs

25 black peppercorns

Directions

1. Preheat a convection oven to 400°F (204°C) with the fan on high.

2. Heat the oil in a large roasting pan over high heat. Add the onions, carrots, and celery and sauté until they caramelize, about 12 minutes. Add the tomato paste and sauté, about 3 minutes. Add the red wine and reduce by half, about 20 minutes. Set the wine and vegetable mixture aside.

3. Meanwhile, spread the chicken wings in a single layer on two large rimmed baking sheets and roast until golden brown, about 50 minutes, turning after 25 minutes. Drain and discard any rendered fat.

4. Transfer the roasted wings to a large stockpot and add the chicken feet and water. Bring to a simmer over medium heat and skim the stock of all impurities and fats that rise to the top.

5. Add the wine and vegetable mixture to the pot, along with the bay leaves, thyme sprigs, and peppercorns. Simmer the stock over low heat, uncovered, for 6 hours, skimming every 30 minutes.

6. Strain the stock through a chinois, return it to the pot, and reduce over low heat to 4 cups.

7. Prepare an ice bath. Strain the reduced jus through a chinois and chill over the ice bath. Store in an airtight container refrigerated for up to 3 days or frozen for up to 30 days.

Citus Gastrique

Ingredients

1 star anise
400 g red wine vinegar
400 g sugar
Finely grated zest and juice of 6 lemons
Finely grated zest and juice of 6 limes
Finely grated zest and juice of 6 oranges

Directions

1. In a sauté pan, toast the star anise over medium heat until fragrant, about 2 minutes. Add the red wine vinegar to the star anise and keep warm.

2. Cook the sugar in a dry saucepan over medium heat, swirling to caramelize evenly. When the sugar is deeply caramelized, add the star anise and red wine vinegar, whisking to fully incorporate, and reduce by half, about 20 minutes.

3. Add the citrus juices and reduce by half again, about 30 minutes.

4. Mix in the citrus zest and let cool to room temperature. Remove and discard the star anise. ⚛

Mixing Texture and Flavor

Recipes mix texture molecules and flavor molecules. The number of flavor molecules in a food is important because we can taste something only if there are enough molecules. But molecular interactions are also responsible for texture. The way this comes about is often quite complicated, and we will spend most of the book examining this in different examples. As a start, let's think about the wildly different *numbers* of texture molecules that conspire together in food.

COUNTING THE TEXTURE MOLECULES IN A COOKIE

Imagine that you wanted to draw the molecular structure of a chocolate chip cookie. You'd have strands of starch, molecules of salt and sugar, larger fat molecules, and so on. The picture would look very different if there were, say, many more long starch molecules than little sugar molecules. In other words, knowing the relative numbers of each molecule changes the picture. It's obvious that the cookie dough contains more grams of sugar than salt, but a molecule of salt is much smaller than a molecule of sugar, so how do the numbers compare? To find out, let's rewrite the recipe. Instead of listing the volume or the weight of the ingredients, let's list the numbers of molecules.

Cookie dough is just a combination of all of the molecules of its ingredients. To find the number of molecules of a given ingredient, we simply take the weight of the ingredient and divide by the weight of an individual molecule of that ingredient. The table below shows these numbers. In the first row we have baking soda, or sodium bicarbonate, which consists of a single sodium, two carbons, and three oxygens. The weight of a single baking soda molecule is 1.4×10^{-22} grams (you get this number by looking up the molecular weight of baking soda, which is 84 g/mol and dividing by the number of molecules in a mole, 6×10^{23} molecules). The recipe calls for 1 teaspoon of baking soda, which corresponds to approximately 5 grams. This means that the total number of molecules of baking soda is 5 grams/1.4×10^{-22} grams $= 3.6 \times 10^{22}$, or 3.6 followed by 22 zeros.

For the other rows in the table, let's proceed with similar calculations: Salt has a single sodium and a single chloride, and thus each salt molecule has a weight of 9.6×10^{-23} grams (or a molecular weight of 58 g/mol). This gives 6×10^{22} molecules. The story gets a bit more interesting when we consider the carbohydrates, proteins, and fats in the recipe, as these come from many different sources. For example, carbohydrates come from sugars (both granulated and brown), but also from the flour. Flour consists of both starch granules, which swell up when cooked, and proteins like gluten, which give the dough its strength and stretchy character. We will consider both of these molecules in more detail later on, but for now we just want to count how many there are in the cookie recipe, thus giving us the total number of carbohydrates. Now let's move on to the proteins. There are two main sources of proteins: the eggs and the gluten in the flour. There are many different proteins in eggs, the dominant one being ovalbumin. Fats come from both the butter and the egg yolks, in the form of various fatty acids.

Putting the numbers together, we arrive at the totals shown in the table. All in all, there are about the same number of fat and sugar molecules (10^{23}). By comparison there are about one-tenth the number of baking soda and salt molecules, and an even smaller number of protein molecules (10^{20}). Interestingly, there is the smallest number of molecules of starch and vanillin—of course, starch is huge and vanillin is tiny. The most common molecule is water (10^{24}).

Therefore, if you look at the mush of cookie dough at the molecular scale, you will see lots of large fat and carbohydrate molecules, surrounded by about the

Ingredient	Total weight in recipe	Weight per mole (weight for 6×10^{23} molecules)	Number of molecules
Baking soda	5 g	84 g/mol	3.6×10^{22}
Salt	6 g	58 g/mol	6×10^{22}
Vanilla[1]	0.02 g	152 g/mol	8×10^{19}
Carbohydrates:			
Granulated sugar	150 g	342 g/mol	2.6×10^{23}
Brown sugar	150 g	342 g/mol	2.6×10^{23}
Starch (flour)[2]	202 g (75% of flour)	60,000,000 g/mol	2×10^{18}
Proteins:			
Eggs[3]	10 g +	50,000 g/mol	4.5×10^{20}
Flour (glutenin and gliadin)[4]	27 g (10% of flour)		
Fats[5]:			
Butter	160 g +	270 g/mol	3.8×10^{23}
Eggs	10 g		
Water:			
Butter	65 g +	18 g/mol	4.7×10^{24}
Eggs	75 g		

[1] Less than 0.5% of the vanilla extract is made of vanillin molecules.

[2] The starch in flour consists of amylose and amylopectin, whose molecular weights vary depending on how many glucose monomers they consist of, ranging from 500 to a couple million. Here we assume a third of a million monomers, which gives a total molecular weight of 60 million g/mol.

[3] Eggs contain many proteins. Ovalbumin, which has a molecular weight of 43,000 g/mol, is the most common. On average we can estimate the molecular weight of the proteins to be about 50,000 g/mol.

[4] For the purposes of this discussion, we can assume that glutenin and gliadin, the two proteins that make up gluten, have a molecular weight of 50,000 g/mol. This is close to the molecular weight of the proteins of gliadin and in the lower range for glutenin, which can have a molecular weight as high as a couple million g/mol.

[5] Both eggs and butter contain many kinds of fats. Eggs primarily have oleic acid, palmitic acid, and linoleic acid, whereas butter has palmitic acid, myristic acid, oleic acid, and stearic acid. We can estimate the average molecular weight of the fat molecules to be about 270 g/mol.

same number of tinier water molecules. The other molecules (proteins, salt, vanillin) are present in much lower numbers—but be careful to include them, otherwise you risk changing the entire recipe! If you don't believe this, go ahead and remake your cookie recipes without salt or baking soda. We bet you will be much less interested in eating them!

SIDEBAR 6: MEASURING VOLUMES ISN'T VERY ACCURATE

Weight of 1 cup of water as measured by course participants (n = 765)

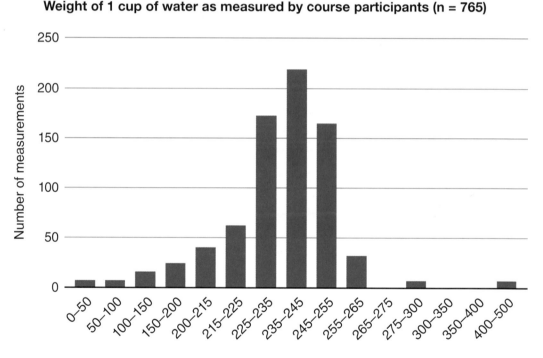

Weight of 1 cup of water (grams)

Careful cooking protocols use weights of ingredients instead of volumes. This is because what actually affects the outcome of a dish is the number of molecules of each ingredient we have added. This is more accurately measured by weight than volume, which can be affected by many things—including how hard you press down on the ingredients in the measuring cup before you weigh it, and even the accuracy of the measuring cup itself.

Are you sure you believe that the 1 cup mark on your measuring cup contains exactly a cup? We don't want to scare you, but our experience is that there are substantial errors.

You can test this at home by taking a measuring cup that holds 237 mL, filling it with water, and measuring the weight (in grams) of the water using a digital scale. Be sure to subtract the weight of the cup by pressing the "zero" or "tare" button after placing the empty cup on it. If you fill an accurate measuring cup to the 237 mL line, the water should weigh 237 g because the density of water is 1 g/ml. If not, it's likely that the lines drawn on your cup are not accurate. (The other possibility is that your scale is not accurate, but this is usually easily remedied by following the manufacturer's calibration instructions.)

This experiment was repeated in the Science and Cooking HarvardX course by thousands of participants. The table to the left below shows the spread of weights that people measured. As you can see, volume measurements are not always accurate, so 237 mL for the author of a recipe may not be the same as 237 mL in your kitchen. To further complicate matters, 237 mL is just about equal to "1 cup" as used in cookbooks in the United States, but in other countries a standard "cup" can have a slightly different volume. Recipes for homestyle cooking (for example, marinara sauce or even chocolate chip cookies) tend to work regardless of precision because they are very forgiving of fluctuations in temperatures and relative amounts of ingredients. However, the same cannot be said for many baking recipes, which rely on precise ratios, nor for some of the chefs' creations that we will see in this book. Part of their magic is that these chefs have learned how to precisely control complex recipes to achieve consistently excellent results. ⚛

MIXING THE INGREDIENTS IN A COOKIE

Having assembled the ingredients, the next step in the cookie recipe is to mix the ingredients together. This may seem like a simple thing, but some of the magic of cookie baking starts to happen as soon as you mix. The reason for this is simple: the ingredients consist of those that are solid and those that are liquid. As soon as the solid and liquid ingredients combine, they transform each other: the sugar and salt dissolve, and the liquid becomes murky and thicker. The maximum amount of a solid that can dissolve in a liquid is called the *solubility limit*. What is remarkable is that different substances that look very similar can have dramatically different solubility limits.

The most amazing example is salt and sugar. You need both to make cookies. If you were given jars of sugar and salt and asked to tell the difference, you could just taste them. But there is another critical difference: much, much more sugar can dissolve in water than salt. You can try this yourself to see: If you have 1 pound of water, only about a third of a pound of salt can dissolve in it. But a full 2 pounds of sugar can dissolve in that same 1 pound of water.

What does this mean for our cookie recipe? According to the table on page 23, there are about 140 grams of water total in the recipe. Roughly half of this comes from the butter and half from the eggs. All of the salt will easily dissolve, since it is far below the solubility limit. But for the sugar it is a close call: there are 300 grams of sugar, which is about twice the amount of water. When you mix the sugar, it will all barely dissolve! If you put in more sugar, it will separate out and not make a smooth batter. The fat in the recipe also cannot dissolve in the water. The only ingredient that dissolves in the fat is the flavor molecule vanillin, which has fat-like properties that facilitate its blending with the fat molecules.

SIDEBAR 7: JORDI ROCA'S MAGDALENA DE PROUST

In addition to simple recipes such as our cookie recipe, the idea of flavor molecules also plays a critical role in fancier, more complicated recipes. To highlight this, we end this chapter by showing you a remarkable recipe from Jordi Roca, the pastry chef at El Celler de Can Roca in Catalonia, Spain. Jordi wanted to make a dish that tasted like old books from a library. To do this, he needed to isolate the flavor and smell of old books in such a way that he could incorporate it into a dish. His critical idea is that the molecules that taste like an old book are fat soluble and not water soluble. So to get the flavor out of an old book, he took an old book that had a particularly strong "old book" odor, and spread lard on the pages of the book. He put the book in a sealed plastic bag and macerated, or softened, it for 12 hours in a water bath at low temperature (he essentially cooked the book sous vide, a technique we will discuss later on). Then Jordi removed the lard from the book and used distillation to separate out the flavor molecules from the fat. With this flavor isolated, he could readily use it in a recipe.

Magdalena de Proust

Ingredients

Rice paper and edible ink
Madeleine Ice Cream (recipe follows)
Darjeeling Tea Pastry Cream (recipe follows)
Crunchy Filo Pastry (recipe follows)
Old Book Essence (recipe follows)

Directions

1. Print extracts from the book *À la recherche du temps perdu* by Marcel Proust on the rice paper using edible ink.*

2. For the plating, make a quenelle with the madeleine ice cream and place on a plate. Cover with a layer of the Darjeeling tea pastry cream. On top, alternate the crunchy filo pastry sheets and the printed rice paper. Add a drop of the diluted old book essence.

Good luck with this step in your kitchen!

Madeleine Ice Cream

Ingredients

4440 g whole milk
1840 g heavy cream
400 g nonfat milk powder
1700 g dextrose
490 g sugar
70 g CSIM (Cremodan Sim, stabilizer for ice cream)
10 cinnamon sticks
5 g lemon peel
1000 g Darjeeling Tea Madeleines (recipe follows)

Directions

1. In a large pot, heat the milk, cream, milk powder, and dextrose to 104°F (40°C).

2. Add the sugar, CSIM, cinnamon sticks, and lemon peel and bring to 185°F (85°C), stirring continuously with a silicone spatula.

3. Remove the pot from the heat, remove the cinnamon sticks, and add the madeleines. Crush the madeleines into the mixture, cover, and refrigerate until chilled.

4. Pour the cold cream mixture into an ice cream maker and churn. Store at 0°F (−18°C).

Darjeeling Tea Madeleines

Ingredients
300 g unsalted butter
195 g inverted sugar syrup
75 g milk
8 g Darjeeling tea
4 g salt
367 g all-purpose flour
180 g confectioners' sugar
15 g active dry yeast
365 g beaten eggs

Directions
1. In a medium pot, combine the butter, inverted sugar syrup, milk, Darjeeling tea, and salt. Heat until the butter is melted. Strain into a clean bowl and set aside.

2. Sift together the flour, sugar, and yeast.

3. In a Thermomix, temper the eggs to 86°F (30°C) and emulsify with the butter mixture. Incorporate the flour mixture until the dough is homogeneous.

4. Pour the dough into small shell-shaped silicone molds. Bake without humidity at 350°F (177°C) for 4 minutes. Rotate the molds and bake for 3 more minutes to ensure uniform cooking. Remove from the oven and allow to cool, then unmold and serve or reserve in an airtight container.

Darjeeling Tea Pastry Cream

Ingredients
100 g egg yolks
45 g cornstarch
250 g cold milk
250 g heavy cream
50 g sugar
20 g Darjeeling tea leaves

Directions
1. In a medium bowl, whisk together the egg yolks, cornstarch, and a small portion of the cold milk until smooth. Strain and set aside.

2. In a small pot, combine the remaining milk, cream, sugar, and tea. Bring to a boil, then remove from the heat. Cover and let stand for 20 minutes while the tea infuses the mix.

3. After the tea is infused, strain the cream mixture into a clean pot and add the egg yolk mixture. Bring to a boil, then remove from the heat.

4. Pour the cream into a container. Cover with plastic wrap, and press the plastic down on the surface of the cream, leaving no space for air. Refrigerate.

Crunchy Filo Pastry

Ingredients
1 sheet filo pastry
80 g unsalted butter, melted
50 g confectioners' sugar

Directions
1. Preheat the oven to 400°F (204°C).

2. Use a brush to paint the sheet of filo pastry with the melted butter, place it on a rimmed baking sheet, and sprinkle the confectioners' sugar on top.

3. Bake for 12 minutes, or until golden. Break the crunchy filo into pieces and store in an airtight container.

Old Book Essence

Ingredients
4 kg old books (pages without color ink only)
8 kg deodorized pork fat or lard
Food-grade ethanol

Directions
1. Heat the lard to 104°F (40°C) to melt it, then grease the old books and pages with it. Place in a plastic bag, seal, and macerate in a water bath for 6 hours.

2. Separate the lard from the books. While still hot, mix the lard with an equal amount of ethanol. (The alcohol helps extract the flavor from the fat.)

3. Allow to stand for 6 hours, then filter and distill at 99°F (37°C), pressure = 0.9 bar, to decant the alcohol and obtain the essence of old books.

4. Preserve the distillate in the fridge in an airtight and opaque container. When ready to use, dilute the essence 1:8 (5 mL of essence in 40 mL of odor-free oil). ✾

CHAPTER 2

Heat

I like to think of heat as an ingredient in the kitchen. You take some raw vegetables, or meats, or dough and you add heat and you have cooked foods, which smell and taste and feel very different from the original stuff. The thing about heat is that of all the ingredients in the kitchen, it's the most common. It's also the most mysterious. And that's because it's hard to measure and hard to control. It's invisible. It's not material, the way water or flour is.

—Harold McGee

The *Oxford English Dictionary* defines *cooking* as "preparing food using heat." Heat can be applied in many different ways, including boiling, searing, grilling, or baking. As scientists who like to recast normal English sentences into precise pictures, here is a simple sketch:

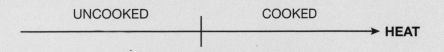

The figure shows temperature on the horizontal axis; somewhere along this line there is a transition temperature, above which the food cooks. Transitions in cooking usually happen abruptly: as soon as it is heated past a critical temperature, the food transforms and is ready to eat.

But what does the heat actually do? What is the difference between food that is cooked and food that is uncooked? In a nutshell, heat causes the ingredients of food to fall apart and then to come back together in different ways. We are used to heat causing things to fall apart in our daily life—fire causes wood to disintegrate, and the energy so released allows the fire to keep burning. The same goes with gasoline in a car: after ignition, the gas molecules fall apart and the energy released allows the car to move. In general, we call these transitions *phase transformations*, indicating that the substance completely changes form.

Although the application of heat during cooking causes food molecules to fall apart, the temperatures used during cooking are vastly lower than those used for burning wood or gasoline. Indeed, putting food directly into a fire helps the fire burn, but hardly leaves an edible product. Cooking is based on the fact that lower heat applied in a controlled way has the ability to transform food into the edible delights that we know and love. This happens because the food ingredients fall apart in stages and then reform in different ways. Every additional increase in heat is crucial, for it causes molecular events that transform the food into the final cooked dish. What exactly happens—and how—depends on the type of molecule and the way that heat is applied. The seemingly infinite number of variations is why we find cooking so multifaceted and interesting.

SIDEBAR 1: CALIBRATE YOUR OVEN

Bakers tend to be very picky about their ovens, and you often hear them say it's because they "know their oven best." The reason for this is that an oven doesn't necessarily bake at the exact temperature it is set to—in fact, most ovens fluctuate quite a bit. This is why

A

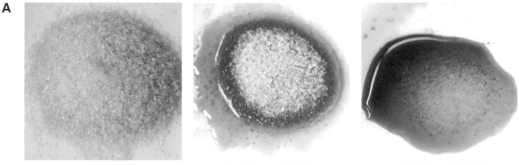

180°C = 356°F   185°C = 365°F   190°C = 374°F

B

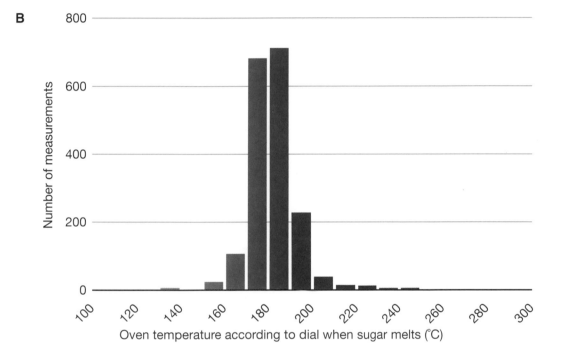

Oven temperature according to dial when sugar melts (°C)

Reported dial settings at oven temperatures of 186°C as
measured by the melting point of sugar (*n*=1881)

you might bake that perfect golden-brown pie crust at home, but at Grandma's house it comes out burnt.

To calibrate an oven, you can of course simply use an oven thermometer—but how do you know if the thermometer is accurate? Another method is to find a material that under-goes a change at some critical temperature and then tune your oven near the critical temperature where the transition occurs to find out if it actually happens at the expected oven temperature. For instance, in *Cooking for Geeks*, Jeff Potter outlines a protocol for calibrating your oven using sugar.

The sugar calibration is based on the idea that the melting point of table sugar (sucrose) is 367°F (186°C). If you put the temperature just below 367°F (186°C), the sugar should not melt. If you raise the temperature just above 367°F (186°C), then the sugar should melt. Try calibrating your own oven using this protocol:

Directions

1. Preheat your oven to 350°F (177°C).

2. Put a teaspoon or two of granulated sugar in an oven-safe dish or on a sheet of aluminum foil.

3. Place the dish in the oven for 15 minutes.

4. If the sugar doesn't melt, raise the temperature of the oven by 10 degrees Fahrenheit (5 degrees Celsius) and repeat the experiment. If the sugar does melt, remove the sugar and lower the temperature of the oven by 10 degrees Fahrenheit (5 degrees Celsius) and repeat the experiment with a new sample of granulated sugar.

5. Continue raising (or lowering) the temperature until you find the lowest temperature setting at which your sugar melts. We will call this temperature T_{melt}. If you had to raise the temperature, this is the final temperature that you tested (the first temperature at which the sugar melted). If you had to lower the temperature, this is the second-to-last temperature that you tested (the last temperature at which the sugar melted).

6. You have thus found that the sugar melts somewhere between ($T_{melt} - 10$) and T_{melt}. So if you find that the sugar melted at 400°F (204°C), but not 390°F (199°C), the melting temperature is between 390°F (199°C) and 400°F (204°C).

Now that you know the approximate "dial temperature" that corresponds to the melting temperature of sugar (367°F/186°C) in your oven, you can calculate the "actual" temperature in your oven for a given dial temperature by simply adding the difference between the temperature you measured for the sugar melting point and the actual melting point. For example, if the sugar melted in your oven at a dial temperature of 355°F

(179°C), your oven temperature is higher than the dial states by 367°F– 355°F = 12°F (186°C – 179°C = 7°C), so you should always set your oven dial 12 degrees Fahrenheit (7 degrees Celsius) lower than what a recipe calls for. If the sugar melted in your oven at a dial temperature of 380°F (193°C), then your oven temperature is lower than the dial states by 380°F – 367°F = 13°F (193°C – 186°C = 7°C), so you should always set your oven dial 13 degrees Fahrenheit (7 degrees Celsius) higher than what a recipe calls for.

Using this protocol, participants in the *Science and Cooking* online course reported their oven temperatures from their kitchens around the world. As you can see in the figure on page 35, most of them were within 10 degrees Celsius of the expected temperature, but a good number of them ran significantly hotter or colder. ✼

SIDEBAR 2: TEMPERATURE, HEAT, AND CALORIES

We often think of temperature and heat as being the same thing—or at least, very closely related concepts. On a 95°F (35°C) day, we would probably say that it is hot; on a 32°F (0°C) day, we would say that it is cold, or lacking heat. Temperature is used to define how hot or cold something is. But what does it actually mean for a substance to be hot or cold? We know that temperature measures the energy in the motion of the molecules in a material. When the molecules move more, the temperature is higher. When they move less, the temperature is colder. Meanwhile, heat refers to the amount of energy that is added or removed from a system. Thus, applying heat makes the molecules move more, which means that the measured temperature goes up. Cooling something down, or removing heat, decreases the motion of the molecules and thus the temperature.

Surprisingly, these ideas—of molecules moving and temperature and heat—are intimately related to calories. You know, those things that make you gain weight when you eat too much of them. Why is there a relationship? Calories measure the energy content of the food. We nominally use this energy to live and go about our daily activities. If we eat food with more energy than we need, then some of it gets converted to extra weight on our bodies.

So how do calories relate to energy or temperature? Calories are units of energy. By definition, a calorie (with a lowercase "c") is a measure of how much energy it takes to heat 1 gram of water by 1 degree Celsius. The number on our food labels is given in

Calories (with a capital C). This is the energy it takes to heat 1 *kilo*gram of water by 1 degree Celsius. In fact, the most accurate way to determine the calorie content of food is simply to burn it, and use the resulting energy to heat up some water! A device called a bomb calorimeter is essentially two connected chambers, one with the burning object and one with water that is being heated up. By measuring the change in the water's temperature, we can determine exactly how much is being released by the burned food.

Let's dig further. Where does the energy in a Calorie come from? You guessed it, the molecules of the food. When we burn a certain food, we are breaking down the bonds in the molecules into their constituent atoms. These atoms then re-form into very stable substances—typically water and carbon dioxide. There is extra energy left over from this process, and this energy is then used to make the molecules move more; that is, it is released as heat. Our bodies do something quite similar when they break down food, though, of course, we manage to break down the food in our stomachs without boiling it. This happens through the action of molecules called enzymes that we will discuss later on. ⚛

Phase Transformations

Phase transformations are all around us in our daily life: Water boils at 100°C (212°F), changing from a liquid to a gas. Water freezes when the temperature drops below 0°C (32°F), changing from a liquid to a solid. These transitions of water are central to cooking, since the most common ingredient in food is water.

More generally, the temperature at which a phase transformation happens depends on the material. Other liquids that look like water behave differently. Ethanol, the active ingredient in alcoholic beverages, changes from a liquid to a gas at a much lower temperature, around 78°C (173°F). So if you heat wine on the stove, the ethanol boils off first, reducing the alcohol content, while only the flavor remains. (This is why you don't become tipsy from the wine you add to a simmering sauce.)

But phase transformations in cooking are much richer and more varied than those of these simple ingredients. For example, consider the case of the egg.

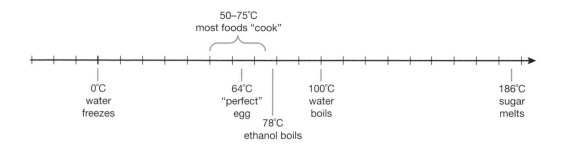

FIGURE 1 If we think about the various transitions that we encounter while preparing food, we find that they all occur within a relatively small range of temperatures—between –20°C (–4°F) and about 190°C (374°F). We store raw ingredients and leftovers in the fridge or freezer because microbes grow extremely slowly or not at all below 4°C (99°F). We cook food with heat both to kill harmful microbes and to denature, or break down, proteins so that the ingredients take on new textures, usually becoming more solid; this all happens between 50°C (122°F) and 75°C (167°F).

When we cook foods that are mixtures of ingredients, the transition temperatures often change, although they remain within that same small range. The system overall may even have a transition temperature that is different from the transition temperatures of its components. This is because phase transitions often depend on complex interactions between the atoms or molecules. Take a look at the temperatures indicated for water and an egg. An egg is mostly water, but its critical transition from a liquid yolk to a solid yolk happens at about 64°C (147°F). This is due to the coagulation of its proteins, which begin to unfold at that temperature.

Initially, when produced by a chicken, the interior of an egg is liquid, surrounded by a solid shell. When you put an egg into the freezer, the liquid transforms to a solid, just as with water. When water is heated, it changes into a gas. But when we heat an egg, something entirely different happens: it transforms into a solid. Moreover, once an egg is heated and hence "cooked," it can't be uncooked. Take a hard-boiled egg and leave it on the countertop at the same temperature at which it started out, and it will remain a solid forever. In contrast, like ice, a frozen egg easily transforms back into a liquid when the temperature rises. How can this be? How is it that the phase behavior of an egg, which is primarily composed of water, can be so different from the phase behavior of water itself? The reason, as we will see, is because of the special molecules inside.

For another example, recall our cookie recipe from chapter 1. We left you with cookie dough, a (tasty) mush of ingredients, but very far from a finished cookie. Remarkably, when the dough is heated, the raw ingredients change beyond recognition. As much as you might love cookie dough, you have to admit that the baked cookie is an entirely different substance. It doesn't flow like a liquid, and its color is completely different. The texture has changed, as it is now full of air bubbles. How did this happen? These types of dramatic changes occur all throughout cooking.

To unravel exactly how heating changes food, let's examine the major food components and their unique responses to heat. Together, their behavior determines what happens to the food overall.

Water

It's tempting to think of water merely as filler for all the other nutritious and flavorful food components: proteins, fats, vitamins, and flavor molecules. But nothing could be further from the truth. On the contrary, water is perhaps the key determinant of what happens to food when we heat it. The reason, of course, is that many foods are mainly water—think of vegetables and fruits, which often contain 85% water or more. Even in a steak, the water content easily makes up three-quarters by weight.

When cooking with heat, the abundance of water in food becomes especially significant. There are two main reasons for this: First, when proteins, carbohydrates, and fats are heated, their response to heat is influenced by the fact that they are surrounded by water. For example, carbohydrates dissolve easily in water. The water will hydrate and change the carbohydrates, but the opposite is also true: the carbohydrates can change how water itself behaves in the presence of heat (we will see this in the next section when we discuss recipes for pralines and ice cream). Like carbohydrates, proteins also dissolve in water, but sometimes, due to heat and chemical properties, they are no longer soluble and form solid bits in the water. Cheese as we know it would not exist without this important fact, so we should pay extra attention to the water-protein interaction. Fats, by comparison, do not dissolve in water at all and tend to exist away from water in most foods. But this too is important, because flavor compounds that are created from heat and are not soluble in water will instead dissolve in fat, resulting in delicious food.

The second consequence of water's prevalence in certain foods is that when heated, the food will behave similarly to water when it is heated: it will boil at around 100°C (212°F). So unless somehow the water goes away, the temperature of the food can't go above the boiling point of water. You can crank up the temperature in your oven all you want, but any water in the food will simply continue to boil until the liquid water vaporizes into gaseous water. But as long as it is in liquid form, the temperature of the food will stay at 100°C (212°F). One tricky consequence is that if you want any transformations in food that require temperatures above 100°C (212°F), you're in trouble. They simply won't take place. You can forget about delicious reactions that need temperatures as high as 150°C (302°F).

There is a way around this, and the secret is to let enough water boil off for these precious transformations to take place. In chapter 4 we will see just how delicate this approach can be. But for now, we hope to have convinced you that water plays an important role when cooking with heat. But we're about to see that it gets even more complicated than this.

SIDEBAR 3:
JOANNE CHANG'S ALMOND PRALINES

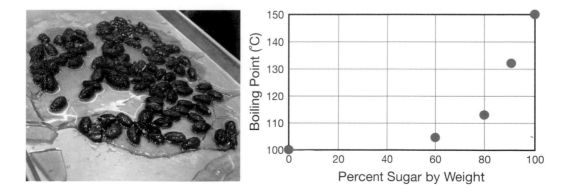

Almond Pralines

Ingredients
2 cups sugar

⅓ cup water

1 cup toasted whole almonds

Directions

1. Combine the sugar and water in a medium saucepan. Bring the mixture to a boil over high heat and let boil, undisturbed, until the sugar starts to caramelize, about 5 minutes. When it starts to color, gently swirl the pan to even out the caramelization.

2. When the sugar is deep golden brown, add the almonds all at once. Give the pan a strong swirl to coat the almonds with the caramel and immediately pour them out onto a rimmed baking sheet.

3. Let cool, then break into pieces and enjoy!

Pralines are made with nuts and caramelized sugar. They are named after the French Marshal du Plessis-Praslin of the early seventeenth century. Stories about their origin vary, but even if it was simply a happy accident of almonds falling into a pot of caramel, they are a remarkable invention that is as scientifically interesting as it is delicious.

Sugar caramelizes at about 170°C (338°F). Pure water has a boiling point of 100°C (212°F), but when sugar molecules are added to the mix, the boiling point of the solution increases. There is a certain amount of pressure on everything due to the atmosphere. To boil, the temperature of the water molecules must increase enough to break the bonds between water molecules and overcome this atmospheric pressure. Above the boiling point, the water molecules enter their gaseous form, which is water vapor. However, the sugar will not evaporate at these temperatures. Normally, we would think of water as diluting the sugar. Here, it helps to flip it around and think about the nonevaporating sugar molecules taking up space and diluting the water molecules. The water molecules have to surround the sugar to make it soluble in the water, and this makes it more difficult for the water molecules to evaporate, resulting in the higher temperature required to transform the liquid water into water vapor.

From the graph on the previous page, we can see that the boiling point reaches 150°C (302°F) when the weights of the sugar and the water in the solution are equal. This means that a 1:1 solution would not caramelize when it boils. The praline recipe, however, has a sugar content that is almost 5 times the amount of water by weight. Since the boiling point increases at a faster rate as the proportion of sugar increases, we can guess that the boiling point of the 5:1 mixture will be much higher than the 170°C (338°F) necessary. This makes it very quick to achieve the golden brown we want, but it is also the reason that it is easy to burn the caramel in the process. If we start out with a smaller ratio of sugar to water and a lot of patience, we could still eventually reach the required temperature and concentration by letting enough water boil off. Conveniently, the fact that we can draw a nice curved line through these points on the graph means that just by measuring the temperature of our boiling sugar water solution, we can know exactly how much water is left in our solution. ❀

Although we know that water in liquid form can't get above 100°C (212°F), let's take another look at Joanne Chang's recipe for pralines. Joanne is the pastry chef of the immensely popular Flour Bakery + Cafe in Boston. Once an applied math major at Harvard, she now regularly returns to speak to our class about the science of sugar. She has made her recipe for pralines in our lecture hall many times. The recipe, a classic take on the candy, could hardly be simpler, at least on the face of it. The sugar and water are dissolved and heated until the sugar caramelizes and darkens in color. The caramelization breaks down the sugar so that it takes on a complex nutty flavor and turns brown. This pairs well with the almonds, which themselves add even more flavor through their browning and toasting process. The result is a wonderfully rich and sweet treat. Remember that the only ingredients are water and sugar, which are combined in a saucepan and heated to a boil. Now, if there was only water in the pan, you know it would boil at 100°C (212°F). But with sugar, something remarkable happens. As the sugar solution keeps boiling, the temperature goes higher and higher. Eventually it can get as hot as 150°C (302°F).

How can this be? For starters, you'll notice that we have a lot more sugar than water in this recipe. We start out with 2 cups of sugar (about 400 g) and only ⅓ cup of water (about 80 g). In other words, the sugar concentration is 83% by weight of the total mixture. Since it's a mixture of two substances, you could imagine that the boiling temperature would reflect the properties of *both* the sugar and water, so this might not seem completely outrageous. But what about the increase in temperature as the solution continues to boil? Let's think about what happens right at boiling. Water molecules are light, about twenty times lighter than sugar molecules, so they boil off easily while the heavy sugar molecules stay behind. As water molecules vaporize and leave the solution, the sugar concentration increases. The solution then takes on even more of the properties of sugar, and the boiling point increases a bit more. This in turn allows for even more water to boil off, which increases the boiling point further, and so on and so forth. This is the feedback loop that eventually results in the scorching temperature of 150°C.

This is a delicious feedback loop indeed because, as it turns out, it is the basis of all candy making. By carefully controlling the sugar concentration, we can modulate the boiling point and create entirely new materials. Joanne Chang's praline recipe shows just how this happens. At the final temperature, the caramelized sugar flows like honey, and when cooled off, it solidifies into a hard but brittle glass-like substance. These textural changes are key to the recipe, and we will consider how they come about in chapter 5. Moreover, the high-temperature sugar solution also develops an intense flavor that is quite different from sugar or rock candy. This is the flavor of caramel. The temperature gets high enough that the sugar molecules break down and produce flavor molecules, something we would not have been able to accomplish if the temperature had never exceeded 100°C (212°F). The flavor molecules are produced through a set of chemical reactions called *caramelization*. A key feature of these reactions is that they occur only above approximately 165°C (329°F), as detailed in the sidebar. The praline recipe is remarkable given the numerous transitions in a recipe with only three ingredients! Note, though, that by the time the recipe is finished, there are actually many more components than just sugar, water, and almonds—these are the myriad flavor molecules that are produced by heating.

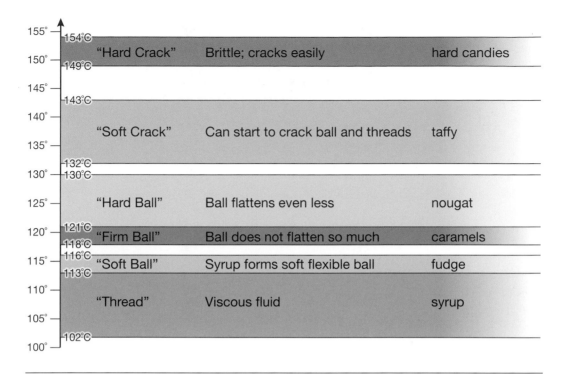

FIGURE 2 Sugar-water solutions are the basis for many confections. In the past, candy was made by hand at home or in small shops. As technology developed, the process became more precise, and new methods were discovered and new candies invented. Nowadays, we know exactly which temperatures are necessary for each type of candy and we have the tools to precisely control the process. Since the temperature of a boiling sugar-water solution is directly related to its concentration, the temperature is used as a marker both industrially and in home cooking. In general, higher sugar concentrations are used to make hard candy such as lollipops, and lower concentrations are used to make softer candy, like fudge or caramels.

The stages of a sugar solution and their corresponding temperature ranges are shown in the figure. The name of each stage is quite endearing: thread, soft ball, firm ball, hard ball, soft crack, hard crack. Where do these names come from? Believe it or not, they come from what happens when you plunge a small amount of the solution into cold water: Soft ball feels like a soft ball. Hard crack is hard to crack. Early in the history of candy making, there weren't accurate thermometers, and this "plunge the solution into water" method was in fact how people tested a sugar-water solution. It's a nice experiment to do at home, but be very careful as these solutions are extremely hot! The thread stage forms threads, as when honey is poured into a cup of water; the syrup at this stage contains so much water that it is still a viscous fluid. The ball stages form small clumps, and the words *soft*, *firm*, and *hard* refer to the stiffness of the clump. At the crack stages, the sugar again forms threads, but they are solid and make an audible cracking noise when broken. The hard crack stage is the most brittle one. Not far beyond hard crack, all the water will have boiled off and the sugar will begin to caramelize. What changes between these different regimes is how much water remains inside the sugar mixture. As you increase the temperature, the water content drops precipitously. Candy making is *extremely* sensitive to changes in temperature. If you make a recipe and stop at the wrong stage, it will make a big difference in the outcome. This isn't so surprising when you realize that only a few degrees separate hard ball from soft ball.

SUGAR + HEAT = CARAMELIZATION

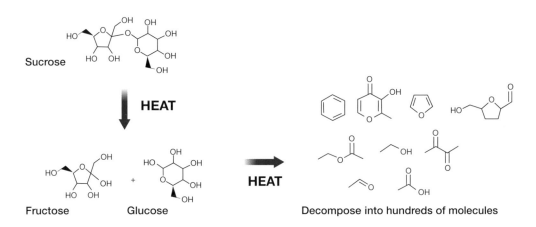

What happened to the molecules in Joanne Chang's praline recipe? This figure shows the molecular detail: caramelization reactions produce many flavor molecules using energy from heat. The first step is that sucrose breaks down into its two components: fructose and glucose. From here the molecules fall apart further into hundreds of molecules. Some of these molecules are brown and change the color from white, and some of them have delicious nutty, toasty, caramel flavors. ⚛

Heat is also used to create texture and flavor out of sugar in other cooking, not just candy making. In chapter 5 we will show a recipe for *crema catalana* from Carme Ruscalleda, chef of the famed Sant Pau restaurant outside Barcelona. The final step of the recipe creates a thick crust of caramelized sugar on top of a cream base. The crisp top layer is produced with a hot iron after the top has been sprinkled with sugar. As with the pralines, heat caused a dramatic change in both texture and flavor in this recipe.

Pressure

The praline recipe taught us an important lesson: heating the sugar water increased the boiling point of water. This enabled more sugar to dissolve in the water, which completely changed the texture and flavor of the raw ingredients. Cooking with heat is all about changing texture and flavor.

Generally, we can accomplish this if we can get the temperature of the food well above the boiling point of water. But here we have a real and fundamental problem: the boiling point of water is only 100°C (212°F). If you add stuff to the water, the boiling temperature goes up a little, but not a huge amount. With the praline recipe, we were able to get the temperature to go up much more, but that happened only once most of the water evaporated. For most foods, this is impractical—a steak without water is dried-out and very different from the delicious juicy steak that you might be used to. Fortunately, science provides us with another convenient way of changing the boiling point of water that works even when there is still lots of water left: changing the pressure.

Figure 3 shows a phase diagram of water, but this time as a function of both temperature and pressure. You see that at 1 atmosphere of pressure, like that in Cambridge, Massachusetts, where we are writing this book, the boiling point of water is the promised 100°C (212°F). But when the pressure increases, the boiling point also increases. This happens because increasing the pressure packs the water molecules together so that it requires more heat for them to break apart and fly into a gas. Note that the phase diagram shows that the freezing point of water is not much affected by changing the pressure. This is because solid ice has all of the molecules packed together so tightly that increasing the pressure by a relatively small amount doesn't have that big of an effect.

Chefs use changes in pressure to manipulate their cooking. Devices exist for both raising and lowering the pressure, and each has its own uses.

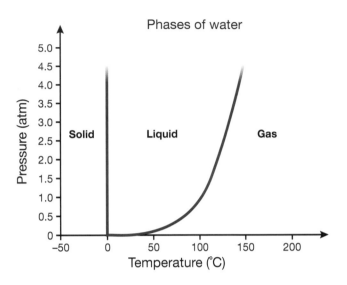

FIGURE 3 This phase diagram for water shows the relationship between temperature, pressure, and the phases (solid, liquid, gas). For any combination of temperature and pressure, we can figure out in which phase we should expect to find the water. For example, the pressure at sea level is 1 atm. At that pressure, the liquid-to-solid transition happens at 0°C (32°F), and the liquid-to-gas transition happens at 100°C (212°F). But at an altitude of 5,000 ft., about the elevation of Denver, Colorado, the pressure is only 0.83 atm. This means that although water will still freeze at 0°C (32°F), the boiling point drops to about 94.5°C (202°F). At this lower temperature, cooking times may need to be increased as much as 50%. But wait! The two axes on this graph are temperature and pressure. By increasing the pressure in our cooking vessel, we can actually overcome the drop in boiling point due to the high altitude. This is exactly the reason why pressure cookers are favored among people who live at high altitudes.

RAISING THE PRESSURE

You can raise pressure in two ways: either dive to the bottom of the ocean to light your fire or use a pressure cooker. For most people, the latter is easier. The pressure cooker was invented in the 1670s by the French mathematician Denis Papin. Nowadays most people use them as a way of accelerating their cooking. The pressure cooker cuts down on cooking time thanks to the boiling point of water: if you operate your pressure cooker at the commonly used pressure of 1.02 atm, the total pressure on the food in the pot will be 2 atm—1 atm from the atmosphere and 1 atm from the pressure cooker. At this pressure the boiling point of water goes up to 120°C (248°F), as seen in the phase diagram above.

But there's another interesting factor: pressure cookers can also be used to create flavor. Recall what we learned from Joanne Chang's praline recipe. Boiling point elevation from heating sugar water raised the temperature enough for caramelization to occur. But as we mentioned before, this happened by getting rid of most of the water. What if we want to make a vegetable soup that tastes like browned vegetables? In typical boiling water, the temperature is too low for browning. And if we brown the vegetables before putting them in the soup, the flavor is dramatically diluted.

By cooking carrots in a pressure cooker, Nathan Myhrvold discovered he could get the temperature high enough to allow for caramelization of the carrots. The result is delicious flavors that would be impossible to obtain with traditional cooking methods.

The quality of the soup depends entirely on the quality of the carrots that go into it, so use the highest-quality carrots that you can find. Carrot cores, although rich in calcium, can add a bitter taste and unpleasant texture to this delicate soup, so you should remove them. It's an optional step, however; you can try the soup both ways and compare.

Caramelized Carrot Soup

Ingredients

500 g carrots, peeled
113 g unsalted butter
30 g water
5 g salt, plus more to taste
2.5 g baking soda
635 g fresh carrot juice
40 g Stovetop Carotene Butter (recipe follows)

Directions

1. Core the carrots by quartering them lengthwise and slicing away any tough or fibrous cores. Cut the cored carrots into pieces 5 cm/2 in long.

2. Melt the unsalted butter in the base of a pressure cooker over medium heat.

3. In a bowl, stir together the water, salt, and baking soda, then add the mixture to the pot, along with the carrots.

4. Pressure-cook at a gauge pressure of 1 bar/15 psi for 20 minutes. Start timing when full pressure is reached.

5. Depressurize the cooker quickly by running tepid water over the rim.

6. Using a countertop blender, carefully blend the mixture to a smooth purée, then pass the purée through a fine-mesh strainer back into the pot.

7. Bring the carrot juice to a boil in a separate pot, then strain it through a fine-mesh strainer. Stir the juice into the purée. Add water, if necessary, to thin the soup to the desired consistency.

8. Using an immersion blender, blend the carotene butter into the soup until it has just melted.

9. Season with additional salt, if necessary, and serve warm.

Stovetop Carotene Butter

Ingredients
700 g fresh carrot juice
450 g unsalted butter, at room temperature

Directions

1. Bring 450 g of the carrot juice to a simmer in a small saucepan.

2. Using an immersion blender, blend the butter into the carrot juice. Simmer for 1½ hours.

3. Remove the pan from the heat and blend in the remaining 250 g carrot juice. Let the mixture cool, then cover and refrigerate it overnight.

4. Warm the carrot butter over medium-low heat until melted.

5. Strain the melted butter through a fine-mesh strainer lined with cheesecloth.

6. Pour the strained carotene butter into molds to set. ⚛

Chef Nathan Myhrvold designed a brilliant recipe for carrot soup that solves this problem. On the surface, the recipe looks like a normal carrot soup recipe, but its taste is much richer and more intense. How does this happen? The carrot soup is made in a pressure cooker that operates at 15 pounds per square inch, or 1.02 atm. Together with the atmospheric pressure, this makes the boiling point of water go up pretty dramatically from 100°C (212°F) to 121°C (250°F).

Because carrots consist of mostly water, the temperature of the carrots also gets up to 120°C (248°F), much higher than the normal boiling point of water. This temperature is high enough for Maillard reactions to begin to occur. These reactions involve both proteins and carbohydrates, which break down into smaller flavor molecules. Some of the molecules have a brown color, as well as an intense flavor that makes the soup much tastier. A second trick Myhrvold uses is to add baking soda, which raises the pH of the soup. It turns out that increasing the pH makes the proteins *want* to give up some of their positively charged H+ ions and interact with the sugars, thereby accelerating the browning reactions. Together, these methods make the richest carrot soup you will ever taste—the carrots are literally browned all the way through! Try it. You'll be surprised at how different the flavor becomes.

SIDEBAR 6: MAILLARD REACTIONS

Nathan Myhrvold's carrot soup was delicious because of the browned carrots. Even if you haven't cooked it, surely you have eaten a piece of golden-brown toast or a perfectly pan-seared steak. All of these foods result from Maillard reactions, which cause the delicious, complex flavors that we associate with the crisp exteriors of these foods. Maillard reactions were discovered in 1912 by Louis-Camille Maillard. He was trying to figure out how biology creates proteins, but instead he stumbled upon what is perhaps the most important reaction in cooking.

The Maillard reaction requires three things: proteins, sugars, and heat. In the first step, an amino acid on a protein—the image shows the amino acid Asparagine—binds to a glucose on a carbohydrate. The resulting molecule continues to break down, bind to new

CARBOHYDRATE + PROTEIN + HEAT= MAILLARD REACTIONS

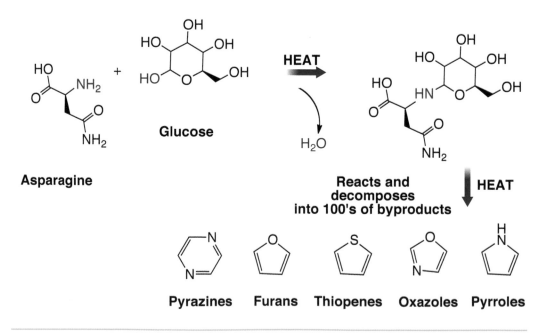

Asparagine   Glucose   Reacts and decomposes into 100's of byproducts

Pyrazines   Furans   Thiopenes   Oxazoles   Pyrroles

molecules, and break down further, in the process forming hundreds of new flavor molecules. The final set of molecules varies depending on the amount and types of protein and sugar content in the food. It takes a lot of energy for all of these reactions to happen. The energy comes in the form of heat, and it has to be pretty hot to get immediate results. The temperature at which water boils (100°C/212°F) isn't nearly hot enough for these reactions; we would ideally like it to be at least 120°C (248°F).

There are a number of ways to increase the amount of browning on your food. One is to increase the pH by adding a base; this works because a basic solution speeds up the first step of the reaction where the amino acid and sugar connect. An extreme example of this is pretzel making, where a very basic lye solution is brushed on the dough to cause that characteristic dark brown color once it has been baked. Another way is to brush your pastry or bread with egg or milk, thereby providing more protein (and, in the case of milk, more sugar) for Maillard reactions to occur. ✿

LOWERING THE PRESSURE

We've seen what happens when you increase pressure, but other magical transformations are possible when you instead lower it. The easiest way to lower the pressure is to climb a mountain and light a fire on top, since pressure in the atmosphere decreases with height. The pressure at the top of Mount Everest is about one-third the pressure in Cambridge, Massachusetts. The phase diagram in Figure 3 shows that this causes the boiling point to decrease. The boiling temperature of water on Mount Everest is about 70°C (158°F), so the recipes that we use at sea level don't work so well. You can let your pasta water boil for a very long time, and the only thing that will happen at such a high altitude is that the water eventually boils off, without the pasta having reached high enough temperatures to cook it the way you are used to at lower elevations. This literally causes everything to change in every recipe you might try, and figuring out how to fix it is not simple.

But sometimes chefs find it advantageous to lower the boiling temperature. An innovation in flavor creation in cooking came about when chefs began using a rotavap (see Figure 4) in their kitchens. Chemists call these devices rotary evaporators, and they have long been a mainstay of chemical laboratories, allowing chemists to separate or concentrate molecules. Some of the first chefs to use a rotavap to manipulate flavors in liquid foods were Ferran Adrià of elBulli and Joan Roca of El Celler de Can Roca in Spain, both of whom you've already met.

Chefs often use a rotavap for concentrating delicate flavor molecules by boiling off water. Of course, boiling off water to concentrate flavor is not unusual in cooking: many sauces, including marinara and wine reduction sauces, are made this way. In these common examples, the boiling water evaporates without damaging the flavor itself, as the ingredients used in the recipes are quite robust. However, there are many subtle cases where the classical reduction method won't work—it will cause flavors to change when heated to 100°C (212°F), since this is above the temperature where many things cook. In addition to concentrating flavors, rotavaps can also be used to extract and separate specific flavors, which can then be used to enhance other foods.

Perhaps the world's greatest master of unusual flavor creation with a rotavap is Jordi Roca, brother of Joan Roca, whose Magdalena de Proust dessert recipe we encountered in chapter 1. He is known for innovative dishes that are designed to stimulate the senses with unusual flavors. For example, his "rainy forest" dessert

FIGURE 4 The photo shows a rotary evaporator, or rotavap, which allows cooking at low pressures. It is a common device in chemistry labs, and has made its way into many haute cuisine kitchens in recent years. Rotavaps work by lowering the pressure so that solutions can boil, and hence evaporate, at lower temperatures than their typical boiling points. The solution of interest is placed in a round glass flask (seen on the right in the photo). The flask is lowered into a water bath that is gently heated. Simultaneously, the pressure is lowered with a vacuum pump. The flask spins continuously so as to create a large surface area from which molecules can evaporate. When the solution starts to boil, the most volatile molecules will turn into gas, and eventually reach the chilled glass coil seen in the top left of the machine in the picture. The chilled glass coil makes the molecules condense back to a liquid and finally collect in the large round glass flask on the lower left. Over time, the solution in the original flask is concentrated and only contains certain non-volatile molecules that have been left behind, whereas the collection flask contains the molecules that have evaporated. Thus the concentrated starting solution and the recondensed liquid have unique flavors, and the decision of which one to use depends on the final dish. The real power of this technique is that since lowering the pressure lowers the boiling point, a rotavap allows us to evaporate water or other compounds without needing to boil the solution at 100°C (212°F), so certain flavor molecules that would be destroyed at that temperature can be preserved. It is also a great way to extract aroma compounds; these compounds would normally dissipate quickly upon heating, but this low-heat process keeps them infused in the water.

extracts the water and smells from dirt to evoke memories of childhood. The extraction is done using a rotavap, which leaves behind the soil and isolates only the much lighter flavor molecules.

A rotavap works by decreasing the boiling point of water, using the fact that the boiling point depends on atmospheric pressure. Why does decreasing the pressure decrease the boiling point? Think about it this way: there is air all around you that squishes your body. The force of the air is larger than you would think—nearly 15 pounds per square inch, so every square inch of your body experiences a pressure of almost 15 pounds. Just imagine dumbbells of this weight pushing down on every square inch all over your body and you get the point that this is not an insignificant amount of pressure. When this pressure is relieved, it is easier for things to boil off. By contrast, increasing this pressure makes it harder for things to boil off. The rotavap works by acting in the low pressure range, letting the flavors boil off and then condensing the vapor by cooling it down, thus creating an intense liquid flavor. Because it can do this at low temperature, the resulting flavor is not cooked but instead tastes fresh. Simultaneously, you can collect the flavors that boiled off, thus allowing you to extract and use them in other creations.

SIDEBAR 7: JORDI ROCA'S NÚVOL DE LLIMONA

Núvol de Llimona

Ingredients

Bergamot Gel (recipe follows)
Lemon Cream (recipe follows)
Lemon Sponge Cake (recipe follows)
Beurre noisette
Lemon Sorbet (recipe follows)
Milk Cloud (recipe follows)
Grated lemon zest
Carnation flowers

Directions

1. On a large plate, make a spiral with the bergamot gel, leaving a 1 cm gap between the lines of the spiral. In that space, draw another spiral with the lemon cream to alternate the ingredients. Set 3 cubes of lemon sponge cake and 1 knob of beurre noisette around the edge of the plate.

2. Serve the lemon sorbet next to it and the milk cloud on top. Grate some lemon zest all over and garnish with carnations of different colors.

Bergamot Gel

Ingredients

500 g water
100 g sugar
8 g agar-agar
100 g lemon juice
Bergamot essence

Directions

1. In a small pot, bring the water and sugar to a boil, then blast-chill it or put it in the freezer for 30 minutes. Add the agar-agar, mix with an immersion blender, and bring to a boil again.

2. Remove the pot from the heat and add the lemon juice and bergamot essence.

3. Homogenize with an immersion blender, transfer to a piping bag, and refrigerate to obtain a gel.

Lemon Cream

Ingredients
500 g Lemon Pith Syrup (recipe follows)
70 g heavy cream
25 g unsalted butter
100 g lemon juice

Directions
1. Remove the piths from the syrup and put them in a Thermomix. Add the cream, butter, and lemon juice and blend for 5 minutes at 149°F (65°C).

2. If necessary, thin the cream with some of the lemon pith syrup to obtain a creamy texture. Transfer to a piping bag and refrigerate.

Lemon Pith Syrup

Ingredients
500 g water
100 g sugar
500 g lemon pith
500 g simple syrup (equal parts sugar and water by weight, brought to a boil and let cool)

Directions
1. In a small pot, bring the water, sugar, and lemon pith to a boil. Remove from the heat and set aside to cool.

2. Repeat four times.

3. Transfer the dry piths to a bowl and add the simple syrup. Leave to stand for 24 hours.

Lemon Sponge Cake

___

Ingredients

200 g unsalted butter

Grated zest of 4 lemons

245 g all-purpose flour

120 g icing sugar

10 g baking powder

3 g salt

50 g milk

245 g eggs

130 g inverted sugar syrup

Directions

1. In a small pot, melt the butter with the grated lemon zest.

2. Transfer to a bowl and add the flour, icing sugar, baking powder, and salt. Mix well.

3. Add the milk, eggs, and inverted sugar syrup and mix well. Cover the dough and refrigerate for 24 hours.

4. Preheat the oven to 350°F (177°C). Butter a baking pan.

5. Transfer the dough to the prepared baking pan and bake for 15 minutes.

6. Leave the sponge cake to cool on a rack, then cut it into 1 cm cubes.

7. Store in an airtight container at room temperature.

Lemon Sorbet

___

Ingredients

100 g Lemon Distillate (recipe follows)

Liquid nitrogen

Directions

1. Put the lemon distillate in a bowl suitable for liquid nitrogen. (Note: This is discussed further in chapter 5.)

2. Whisk the distillate, adding liquid nitrogen slowly until a smooth sorbet forms.

Lemon Distillate

Ingredients
350 g lemon peel
400 g water

Directions
1. Clean the lemon peel well, making sure to remove all the pith to prevent bitterness. Combine the lemon peel and water in a bowl and refrigerate for 6 hours to infuse.

2. Leave the infusion in the rotavap at 113°F (45°C) for 1 hour 45 minutes. (Note: This is where the magic happens. The low pressure extracts flavor.)

3. Reserve the resulting distillate.

Milk Cloud

Ingredients
250 g skim milk

Directions
1. Use an immersion hand blender to thoroughly aerate the skim milk. (Note: This makes it into a foam.)

2. Spoon out the resulting foam and submerge it for a few seconds in a container with liquid nitrogen.

3. Quickly reserve in the freezer. ✿

Fats

Water molecules are tiny, about ten times smaller by volume than fats and proteins. As such, the phase transitions of water are relatively predictable, even considering its interesting behavior at various pressures and with different solutes. With the increasing size and chemical complexity of the larger food components, more complex arrays of phase behaviors can occur. Fats are in some ways in between: they are simple yet have some of the complexity of even larger molecules. Fats are incredibly important for cooking. Often, they are a part of the food itself, as in the fat in a marbled steak or the creaminess in a rich cheese. Other times we add them as an ingredient during the cooking process. We added butter to the chocolate chip cookies earlier in this book, and we might add butter to a frying pan before cooking a steak.

If you think about the role of fats in foods, you'll recall that they frequently undergo some phase transition during the cooking process. Mostly they melt or turn solid. In fact, often the phase transition is the very key to a delicious outcome. For example, you really want the butter in the chocolate chip cookie recipe to melt—first only slightly, so it becomes soft and the other ingredients can mix into it, and then completely once in the oven, so the texture of the cookie becomes just right. Without these transitions, the cookies wouldn't be the same. Similarly, you want the butter in the frying pan to melt so the steak won't stick. And you want the fat in the marbled steak to melt (or "render," as cookbooks say) and make for a juicy and delicious steak. It's quite lucky for us that the phase transitions of fats often happen right around the temperatures at which we cook.

A key characteristic of fat's phase behavior is that different fats have different melting temperatures. Coconut oil goes from solid to liquid at 24°C (75°F). This is just around room temperature, so you'll often find that the coconut oil in your pantry is solid on a cold day and liquid on a hot day. Olive oil, by contrast, is always liquid at room temperature, but with a melting temperature of –6°C (21°F), it starts to turn solid when you put it in the fridge. These are just two examples, but virtually any fat, whether shortening or cocoa butter or canola oil, has a unique melting temperature. As is usually the case in cooking, what we observe with the naked eye can be explained by what happens on a microscopic scale. The melting behavior of fats is no different.

Fat molecules are much larger than water molecules, with most of the atoms arranged in three long carbon chains that are connected at one end. The carbon chains look largely the same—they contain only carbons and hydrogens. But different fats are made up of different fatty acids, and small differences in their structure make all the difference for their phase behavior.

The carbon chains can be saturated (straight) or unsaturated (bent). You can see what this looks like in the figure in the sidebar. It is this geometrical difference that accounts for the phase behavior. Imagine taking two saturated fatty acid chains and trying to push them close together. Since they are both straight, you will be able to align them such that they touch each other more or less along their entire carbon chains. This allows for bonds to form between the carbon chains and they will stick to each other. By contrast, imagine instead the same exercise with two unsaturated fatty acid chains. Since they are bent, you will not be able to pack the chains as closely together, so the bonding interactions will not be as strong. It will be easier to break these molecules apart than the tightly packed molecules because there are fewer bonds holding them together.

SIDEBAR 8: STRUCTURE OF FATS (UNSATURATED AND SATURATED)

Saturated fats get their name from the fact that each carbon is saturated with hydrogen atoms. In general, carbon likes to be bonded to four other atoms at all times. In a saturated fat, each carbon is bonded to two hydrogen atoms and two carbon atoms, each linking one carbon to the next. These single carbon-carbon bonds in the molecule are straight carbon chains; as a result, saturated fats have fatty acid chains that are straight. In contrast, the carbon atoms in unsaturated fats have not been saturated with hydrogen atoms, which means that the carbon is no longer bonded to four other atoms. Carbons

MELTING TEMPERATURE DEPENDS ON PACKING

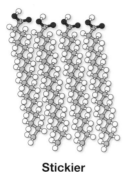 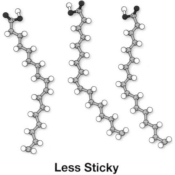

Stickier
Higher melting T

Less Sticky
Lower melting T

with fewer than four bonds are unstable, so instead of binding to hydrogens, some of the carbons form double bonds with each other. Depending on how the double bonds form, they are referred to as either cis- or trans-bonds, and they contribute differently to the overall geometrical structure of the fatty acid. In particular, cis-bonds are notorious for making the fatty acid chains have a bent and uneven structure.

Still following along? If so, here's a question: Which of the fats, saturated or unsaturated, will melt at a lower temperature? The answer is unsaturated fats. They come apart with just a little bit of heat because they are bent and not bonded together as tightly. Saturated fats, on the other hand, require a lot more heat to come apart, which means they have higher melting temperatures. Next question: Which do you think has more saturated fats, olive oil or coconut oil? Recall from above that their melting temperatures are −6°C (21°F) and 24°C (75°F), respectively. The answer is coconut oil. It contains 61% saturated fats compared to olive oil's meager 14%.

But there is more to the melting points of fats than if they are saturated or not. Their carbon chains can also have different lengths. Two very long carbon chains bind more tightly to each other than short ones because they interact with more atoms, all of which can form bonds with other atoms on neighboring carbon chains. Thus, a long chain will require more heat to melt and break all the bonds. You can see now how this begins to get complicated. Fats with very long, saturated carbon chains will have the highest melting points, and fats with short, unsaturated carbon chains will melt at the coldest temperatures. But what about a combination? Will a very long but unsaturated fat have a higher interaction energy than a short and saturated carbon chain? It's hard to tell. It all depends on which property wins out. Luckily, for any given fat, you can always do the experiment in your kitchen to find out. ⚛

We've now learned quite a lot about the phase behavior of fats. It's an important topic because fats are in many ways a middle ground between simple materials, like water, and other complex foods, like eggs, which contain many different types of molecules. Conceptually, they work in much the same way as water. But in more complex fats like cocoa butter, which melts slowly over time, we start to see what happens in materials that contain many different types of molecules. This, of course, is closer to how phase transitions in food work more generally. Cocoa butter is an incredible substance that can be used in many different foods, endowing them with striking properties. Sidebar 9 explains how this is related to the structures and phase behavior of the fat molecules.

PHASES OF CHOCOLATE

A

B

C

18°C 22°C 26°C 29°C 34°C 36°C

Phase 1

Phase V
Best for chocolate

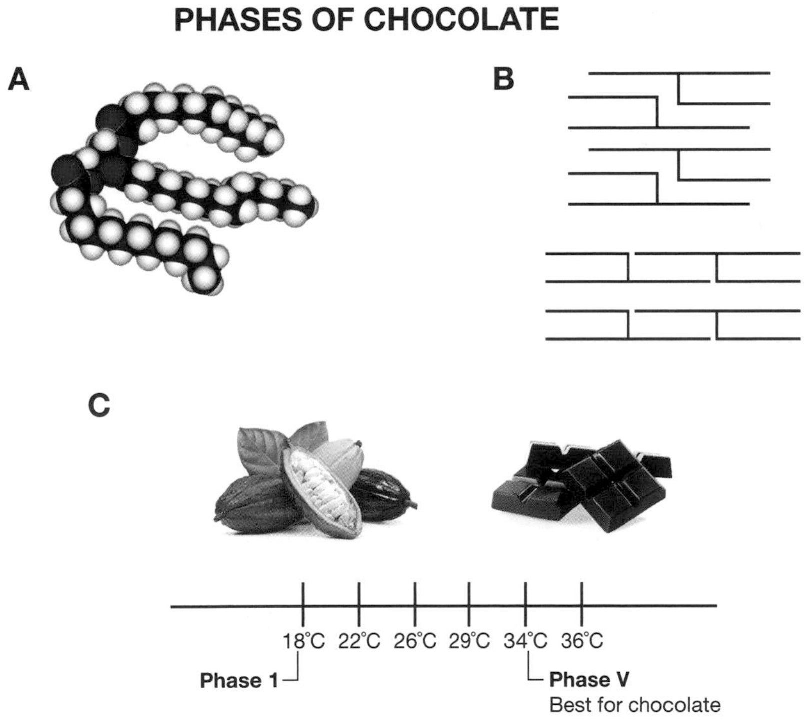

A: Like all fats, cocoa butter is composed of triglyceride molecules. *Tri* refers to the three fatty acids chains and *glyceride* refers to the name of the molecule holding them together. Cocoa butter has many different types of fatty acid chains, each of them with a unique melting point. The fatty acid chains of the molecule can rearrange themselves in different ways. There are six different conformations that we refer to as "phases."

B: The phases of chocolate are due to different crystal structures that correspond to different arrangements of the fatty acid molecules. The shapes of the different arms allow the triglycerides to pack in at least two different ways.

C: As shown in this phase diagram of the different phases of chocolate, there are six different crystalline phases, occurring between 18°C (64°F) and 36°C (97°F). The fifth phase is the one that results in the best chocolate.

Chocolate as we know it today has been around only since the 1800s, but cacao beans have been consumed for thousands of years. Chocolate is made from cacao beans, which contain cocoa solids and cocoa butter. To make bar chocolate, different proportions of the solids and butter are mixed with sugar and milk, depending on the desired sweetness and creaminess. Then, the mixture goes through a process called tempering before it is formed into its final shape.

When cocoa butter is a solid, the fats form crystals, but there are several possible arrangements that the fats can take on. The purpose of tempering is to control the crystallization so that only the desired crystals form.

Crystalline Phase	Melting Temperature	Properties
I	17°C (63°F)	Soft, crumbly, melts too easily
II	21°C (70°F)	Soft, crumbly, melts too easily
III	26°C (78°F)	Firm, poor snap, melts too easily
IV	28°C (82°F)	Firm, good snap, melts too easily
V	34°C (94°F)	Glossy, firm, best snap, melts near body temperature
VI	36°C (97°F)	Hard, takes weeks to form

To make good bar chocolate for eating, crystalline phase V is ideal. Not only does it look aesthetically pleasing (glossy), but it also has a melting temperature that is just below body temperature. This means that it remains solid in most storage conditions, but we can experience the melting sensation once it enters our mouths. Tempering involves heating the chocolate high enough that all the crystals melt, about 49°C (120°F), then cooling the solution down to about 27°C (80°F). Fun fact: a melting temperature also represents a freezing or solidifying temperature. So, the cooling step allows small crystals of types IV and V to begin forming. Then, the mixture is heated slightly to melt the type IV crystals, leaving behind only some type V "seed crystals." At this point, when the chocolate is finally allowed to cool completely, it will follow the lead of the seed crystals and form a uniformly type V piece of chocolate.

You might wonder why we can't just hold the mixture at 34°C (94°F) instead of cooling it down and then heating it a second time. In theory, you could, but the downside is that it would take a very, very long time for the crystals to form. Because temperature is a way to describe the energy and movement of the molecules, crystals do not form the instant you cross the melting/solidifying temperature. Instead, think about that temperature as the point at which the solid and liquid forms are equally likely. The further your temperature is from the transition temperature, the more the molecules will prefer to be in one arrangement over the other and the more quickly the transition will take place.

Chocolate that has not been tempered well, or that has not been properly stored, can result in the separation of some cocoa butter or sugar. If you have ever eaten an old piece of chocolate, it may have had a white layer of "bloom" at the surface—the layer consists of fat and/or sugar. While the chocolate is still safe to eat, it doesn't look very appetizing anymore, and the texture may have become somewhat mushy or gritty. ❀

SIDEBAR 10:
ENRIC ROVIRA'S CHOCOLATE EGGS

Chocolate Eggs

Ingredients
Dark chocolate
Cocoa butter

Directions

1. Choose the molds.

Choose the egg molds of the desired measurements. Make sure that they are of a similar design and that the measurements have proportional increments so that the final composition is harmonious and elegant.

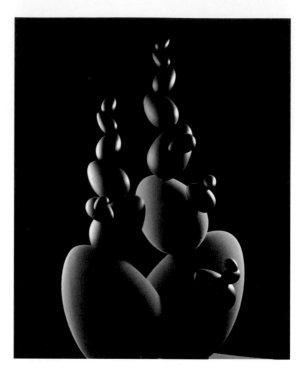

The chocolate molds, if possible, should be made of polished rigid polycarbonate. This material is the most used by professionals and in industry thanks to its poor heat conductivity and high shine, which is transferred to the chocolate. The molds must be perfectly clean, so that the chocolate peels off correctly and the shine is uniform.

The temperature of the chocolate molds should be between 68°F and 77°F (between 20°C and 25°C). If the molds are too cold, the melted chocolate will not disperse correctly, causing air bubbles to appear on the surface. If they are too hot, the crystallization of the chocolate will be slowed down, altering the tempering process, and the result will be incorrect. This will be evident by the chocolate surface not being smooth, lacking shine and exhibiting color changes.

2. Melt the dark chocolate.

Heat the dark chocolate to 113°F to 122°F (45°C to 50°C, according to the manufacturer's recommendations) using a bain-marie or a microwave. At this temperature, which is higher than the melting points of all the crystalline forms of the cocoa butter, all the crystals in the chocolate to be tempered are fully removed.

It is important to carefully monitor the temperature achieved, as a lower temperature than the one recommended will allow some crystals to escape melting, and their presence will affect the viscosity and final crystallization of the chocolate. On the other hand, excessive heating of the chocolate, above 131°F (55°C), can cause the sugars and proteins in the chocolate to start burning, which will change its appearance, flavor, and mouthfeel.

3. Temper the chocolate.

Chocolate tempering is a process in which the melted chocolate crystallizes (solidifies) in a specific way to achieve the desired texture, shine, and melting properties.

The crystalline state of cocoa butter is the determining factor of whether the chocolate is in solid or liquid form. Cocoa butter crystals are polymorphs, which means that they can form in up to six different crystalline configurations, each of which has different physical properties like appearance, density, and melting point.

The process of tempering ensures that the chocolate crystallizes exclusively in phase V, which has the desired characteristics. To achieve this, it is necessary to create, add, and preserve the proper crystallization nuclei in the chocolate.

In order to create the phase V crystallization nuclei, the melted chocolate (free of all crystals) must be cooled down to 82°F (28°C) to facilitate the formation of phase IV crystallization nuclei and then heated back up to 89.5°F (32°C) to promote their transformation into phase V crystals.

Follow these steps to temper chocolate without the need for specialized machinery.

3.1 Melt the chocolate, making sure that all crystals are removed as described above.

3.2 Pour three-quarters of the melted chocolate onto a slightly cold surface (preferably a marble worktop or thick platter) in order to cool it down to 82°F (28°C). Spread and stir the chocolate constantly using a palette knife, triangular spatula, or scraper to make the temperature uniform. The stirring also helps the crystals form; if the chocolate is not stirred, it takes much longer for the crystals to appear. Be careful not to introduce too much air into the chocolate mass. With the correct cooling, crystallization nuclei of the unstable form IV will be created.

The environmental temperature and the temperature and thickness of the marble worktop are of extreme importance in this process. If the marble is too cold, the chocolate will solidify very quickly, and solid chocolate particles will be added to the final mix. If the marble is too thin or not cold enough, the marble will gain heat from the melted chocolate and no cooling will be achieved.

3.3 Combine the cooled chocolate (82°F/28°C) with the remaining one-quarter of the melted chocolate (113°F to 122°F/45°C to 50°C) and mix well in order to ensure a homogeneous temperature. At this point the temperature of the chocolate should be 88°F to 89.5°F (31°C to 32°C), which is the appropriate temperature to allow the IV form crystals to transform into the desired V crystal form.

If the temperature of the mixture is higher than 89.5°F (32°C), cool the chocolate and reheat it again. If it is lower, heat it up to the correct temperature.

3.4   Check the tempering by dipping the tip of a knife or a strip of paper into the chocolate. If the crystallization occurs quickly (in approximately 5 minutes) and the resulting chocolate is smooth and shiny, the tempering process was successful, and the chocolate is ready to be used. In order to work with the tempered chocolate for as long as possible, the achieved temperature should be maintained.

4. Fill the chocolate molds.

4.1   Pour the tempered chocolate into the molds and carefully spread it all over the interior. Lightly shake the molds in order to make sure that the chocolate reaches all corners and to remove any potential air bubbles.

4.2   Empty the excess chocolate out of the molds and allow them to drain upside down on a cooling rack.

4.3   Once the chocolate starts to solidify, but while it is still malleable, cut the excess chocolate off the edges of the molds. Place the molds upside down over the rack once again until the chocolate has set.

4.4   Repeat these steps until the layer of chocolate in the molds is of the desired thickness.

5. Set aside for crystallization.

Tempered chocolate can crystallize at room temperature (64°F to 68°F/18°C to 20°C), but it is best to place the molds (upside down, to facilitate unmolding) at a lower temperature (50°F to 59°F/10°C to 15°C) or in the fridge until the chocolate sets. Complete crystallization (phase V) includes a slight contraction of the chocolate, which facilitates the unmolding process.

6. Assemble the chocolate eggs.

Join the two halves of each egg using a small amount of tempered chocolate or by lightly melting the edges of the eggs to prepare the surface to be joined. In order to do this, slide each of the egg halves over a smooth, slightly hot surface (122°F to 131°F/50°C to 55°C) in order to melt the cocoa butter on the most superficial layer of the edges. Gently press the two halves together and allow the chocolate in the juncture to crystallize completely.

7. Assemble the composition.

Assembly the different-sized eggs using tempered chocolate, maintaining the verticality of all the elements and assuring their stability. Filing the union points with a knife or using a hot surface can facilitate the assembly.

8. Paint the composition.

Prepare a mixture of dark chocolate and cocoa butter (added to reduce the viscosity of the chocolate). Temper this chocolate mix as described above. Using a spray gun, apply a light, uniform layer on the surface of the eggs you want to paint. Use the following instructions to achieve either a satin finish or a velvet finish. Then allow the paint to set completely for a few minutes in a cool environment before further manipulating the composition.

8.1 To achieve a satin finish, the chocolate surface to be painted should be at a temperature of 68°F to 73°F (20°C to 23°C). A lower temperature will produce a grayish matte finish, whereas a higher temperature will slow down the crystallization process, altering the tempering process and creating white spots on the surface (crystallization nuclei of undesirable crystal forms).

8.2 To achieve a velvet finish, you will need to apply the Giner system, a technique developed by pastry chef and chocolatier Joan Giner in 1963. The technique consists of cooling the chocolate pieces to be painted to 39°F to 45°F (4°C to 7°C) before spraying them with the same tempered chocolate preparation used for the traditional painting. By doing this, the sprayed chocolate mix crystallizes quickly as it comes in contact with the cold surface, acquiring a velvety appearance and a lighter color.

The Giner system was studied at the University of Barcelona in collaboration with Enric Rovira. According to the study, the crystallization obtained by using this method is also composed of phase V crystals, but the crystals are smaller and have a slightly lower melting point. These changes give the chocolate a softer texture and increase the sensation of freshness experienced when eating it. ✺

ICE CREAM AND FREEZING POINT DEPRESSION

Joanne Chang's praline recipe showed us how sugar can have a profound effect on the boiling point of water, raising it from 100°C (212°F) to 150°C (302°F) as more and more water boils off. You might wonder: is the freezing point similarly affected? To figure this out, think about what happens if you put water in an ice cube tray in your freezer. The water turns into rock-hard ice cubes. Now, think of the *ice cream* in your freezer. (Surely you have ice cream in your freezer—all reasonable people do.) Ice cream is essentially a sugar-water solution not unlike the one in Joanne's praline recipe. The main difference is cream instead of water, but cream consists of mostly water, so it is still a good comparison. Is the ice cream as hard as the ice? No, it's usually softer. At the temperature in the freezer, water is completely frozen, but water with sugar in it is only partly frozen (and hence softer). This is because the sugar in the water has lowered the freezing point. In order to get it as hard as the ice, we must decrease the temperature of the freezer further. (Notice that you already had all the information necessary to answer the question about freezing point. It's part of why we love asking questions about food and then applying the principles in this book—they can lead to fascinating insights!)

Here's how freezing point depression works on a molecular scale. When water freezes, the moving water molecules are organized into a static and highly structured framework of molecules: ice. The framework forms relatively easily when there are only water molecules in the solution, but sugar molecules are larger and altogether different from water molecules. It's harder to fit them into the framework. The sugar molecules can be made to fit only when they and the water molecules are really slowed down and don't wiggle so much; then the bonds can form that hold them in place. For this to happen, you must lower the temperature. Thus, to make a sugar-water solution freeze, you have to decrease the temperature more than for water. This is the essence of freezing point depression.

Let's look at the recipe for ice cream in the sidebar. You can see that it involves not only large quantities of sugar, but also quite a bit of salt. The first step is to dissolve sugar in cream, and whatever other flavors you want, to make the ice cream mix. Then you put this mixture in a bag with ice and salt, a principle that is not so different from how most ice cream makers work, except that here we're using bags instead of fancy machines with canisters, where similar salt-water solutions are hidden from view.

SIDEBAR 11: ICE CREAM

Ice Cream

Ingredients
600 g ice
200 g coarse salt
100 g whole milk
90 g heavy cream
20 g sugar
¼ teaspoon vanilla extract or other flavor

Directions
1. Combine the ice and salt in a large ziplock bag.

2. Combine the milk, cream, sugar, and vanilla (or other flavor) in a small ziplock bag. Press out as much air as possible and seal the bag, then place it in a second small bag to prevent leaking. Press out the air and seal the second bag.

3. Place the double bag with the ice cream ingredients in the large bag with the ice. Press out the air and seal the large bag. Place the entire package in a second large bag to prevent leaking. Press out the air and seal the second bag.

4. Shake, massage, or gently throw the bag between friends until the ice cream becomes solid, about 10 minutes.

5. Remove the small bag with the ice cream from the large bag of ice. Wipe off the top of the small bag and then open it carefully.

6. Enjoy!

The oldest ice cream makers consisted of two buckets: one filled with the liquid ice cream base placed inside an outer bucket containing salt and ice. By constantly churning (mixing) the cream and sugar mixture as it freezes, we eventually get a batch of cold, creamy ice cream. If more cooling power is needed, we can simply add more ice and salt. These simple machines operate on the principle that we can play around with the phase transitions to get the result that we want. Modern ice cream machines are electrically powered, but the idea is fundamentally the same.

So, what exactly is going on when we make ice cream? The outer bucket contains salt and ice. If you have ever lived in a place where it snows in winter, you have probably seen salt used to melt ice. Adding salt to ice actually lowers the temperature even though the ice is transforming into water. If you're curious to see this in action, a simple experiment you can do is add different amounts of salt to ice water and measure its temperature. When the solution is 20% salt, the temperature should drop to about −17°C (2°F). Since ice cream is a mixture of cream, milk, and sugar and not pure water, it needs to be quite a bit colder than 0°C (32°F) to freeze; the salt helps us get there. The best type of salt for this is a coarse-grained rock salt. Table salt can work, but it is more difficult to evenly distribute the fine grains among ice cubes. They stick and immediately dissolve into the thin layer of water on the ice cube surface, so the temperature will be less even. Some electric ice cream machines do not use salt and ice, but they also rely on a mixture of water and some solute that freezes below 0°C (32°F).

If we merely placed the cream mixture on a cold surface, it would become a hard block of frozen cream. That doesn't sound very appetizing, does it? Instead, we churn the mixture so that the ice crystals in the slowly freezing cream are disrupted as soon as they form. In this way, they remain small and disconnected, which keeps the ice cream smooth and soft. It works as long as there is agitation, and it doesn't matter too much exactly how the crystals are disrupted. In fact, both of these processes can easily be replicated even without an ice cream machine! The recipe above describes a method where ziplock bags are used instead of buckets and the "churning" can be done by shaking or massaging the bag, or even throwing it around—just make sure to use sturdy bags. ⚛

Ice cream is a terrific example of how a scientific concept can inform the making of delicious food—but only if applied correctly. At first glance, the recipe raises a bit of a conundrum: We add sugar to milk in order to make the ice cream sweet, but this lowers the freezing point of the milk, so it's more difficult to make it freeze.

You can put the ice cream in the freezer and just wait for it to freeze, but ideally you should stir while the ice cream freezes so the crystals don't become too large. Instead, you can try to put the ice cream mix in a bucket of ice from the freezer, but the temperature of ice outside the freezer quickly reaches that of water's melting point, 0°C (32°F), and the freezing point of the ice cream mix is lower than this. The solution is to add salt, a lot of salt, to the ice. Salt is just another solute like sugar and has a similar effect when added to water—it lowers the freezing point. In cold-weather climates, salt is frequently poured on the roads in winter as a way to stop rain and snow from freezing. Sprinkling salt on the roads lowers the freezing point enough so that roads and sidewalks don't freeze over and become slippery.

Here's the trick: for the ice cream recipe, you have to add enough salt so the freezing point of the water-salt mixture gets lower than the already lowered freezing point of the ice cream mix. In this case you're lucky because the amount of freezing point depression depends on the number of molecules added to the water. The more molecules, the lower the freezing point. The recipe works because each salt molecule has two ions, one sodium and one chloride, whereas sugar has only one sugar molecule. Salt wins, and you get ice cream—enjoy!

Heating Proteins

We've seen how water, carbohydrates, and fats are important for phase transformations in food. Let's now look at the last of the major food molecules: proteins. Proteins are arguably the most important molecules for life. They are the machines that make your body work: proteins help digest your food, carry signals in your brain, and even help your lungs metabolize oxygen. Every single cell of your body is chock full of proteins, which make the cells divide and grow and do everything needed to keep your body healthy and alive.

When heated too much above the temperature where they are comfortable, proteins stop functioning. Some can be heated up more than others—for example, the proteins from arctic penguins are much less resilient to heat than the proteins inside a desert cactus.

Just as proteins are important for the functioning of living things, the food we eat also comes from living things. This is true whether you are vegan, vegetarian, or a carnivorous farm animal. The vegans among you might not want to hear this, but the first thing that happens when we apply heat to the living things that we eat—plant or animal—is that we kill them. Indeed, death is critically important for food safety. Before you cook it, your food is full of microorganisms—bacteria and fungi— some of which are harmful if eaten. Very small amounts of heat cause the proteins within these microorganisms to malfunction, and the bugs die. If these microbes are not killed, they can make us sick; over time, they can also contribute to spoiling and eventually rotting the food. Most microorganisms live and multiply happily in a temperature range from a few degrees above freezing to about 50°C (122°F). This temperature range explains why food left out at a comfortable room temperature of 25°C (77°F) will spoil relatively quickly. It also explains why bacteria and yeast thrive in our 37°C (98.6°F) bodies, a pleasantly warm temperature if you're a microbe; our body temperature enables the survival of both the good microbes living in our guts and the harmful microbes that can make us sick. At temperatures lower than this range, it is too cold for microbes to survive and grow. This is why freezing and refrigeration are such effective ways to preserve food—the colder the temperature, the slower the microbes can grow, and the longer the food can be preserved.

When the temperature is just a few degrees above 50°C (122°F), the proteins that carry out the essential chemical reactions malfunction, and the organism dies. Why does this happen? Think about proteins for a moment. They are long chains of amino acids that are folded into highly specific three-dimensional structures. When you add heat, the thermal energy causes the protein to start to wiggle. This wiggling causes the protein to be tugged and pulled from all directions; eventually, some of the bonds holding the structure together break, and the protein begins to lose its structure, or unfold. You can see a rendering of this in Figure 5. At first glance, the structural change may be so minuscule that it is not noticeable. But if you were an organism depending on the function of the protein,

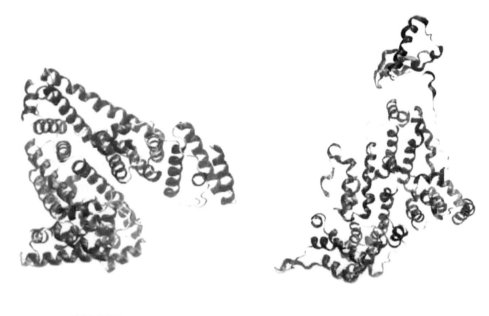

FOLDED   **UNFOLDED**

FIGURE 5 Proteins are very complex molecules. They consist of long chains of amino acids. Even though scientists can relatively easily figure out which amino acids make up which proteins, in many cases the way the proteins fold themselves up is still a mystery. Each amino acid has a different structure, charge, and size. The order of amino acids affects which parts of a protein may fold together; the folding, in turn, affects how a protein functions in the body. The image shows the folded state of egg albumin (left). The egg albumin protein consists mostly of repetitive backbone structures called alpha-helices (shown in purple), and some nonrepeating segments interspersed throughout (shown as thin lines). The image on the left shows the protein in its natural folded state. The image on the right shows a simulation of what happens when you add heat and the protein denatures.

you would certainly notice, since the protein would no longer be able to perform its basic cellular tasks and you would die.

Most proteins start becoming nonfunctional well below temperatures at which we would consider the food "cooked." This makes sense because killing a microbe requires messing up its proteins only a tiny bit, much less than when the food is fully cooked. The tiny, but not insignificant, difference between these two stages—a few slightly unfolded proteins, versus many that are completely unfolded—is something the food industry uses to its advantage all the time when pasteurizing food. Most pasteurization protocols involve adding just enough heat so that the microbes die but the food remains unaltered. This usually means heating the food for different amounts of time. The higher the temperature, the shorter the time. You can, for example, make a perfectly pasteurized egg in your very own kitchen by heating it in a 57°C (134.5°F) water bath for two hours. When you crack open the egg, the yolk and white will look completely raw, but any microbes that may have been present are dead, and the egg is perfectly safe to eat. At higher temperatures the pasteurization is almost instant. For example, the pasteurization protocol for milk as recommended by the National Dairy Council in the United States involves temperatures ranging from 89°C (192°F) for 1 second to 100°C (212°F), the boiling temperature of water, for just 0.01 seconds.

Thus, even with fairly low heat, we can kill microbes and pasteurize food. This is a fantastic, indeed, potentially lifesaving, result—at least for us humans—but, of course, it's not the only reason we cook. What happens when we increase the heat further?

At higher heat, the proteins whose structures were only a little messed up unfold even further. In fact, they change so much that you can see the change with your naked eye. You can't see each individual protein, of course, but you can see that something must have happened—the food changes color and texture! Eggs go from transparent to white, and from liquid to solid. Raw steak turns from pink to brown, and from tender to firm. What is happening is that the proteins are denaturing, or unfolding, and then the individual, separate molecules can stick to one another, causing the proteins to coagulate, changing the liquid into a solid, and cooking the egg.

Cooking Eggs

Let's examine these protein-unfolding transitions in the context of an egg. Eggs are remarkable, the essence of life. Left to its own devices, an egg contains all of the ingredients needed to develop into a live chicken. In terms of cooking, eggs are also remarkable at demonstrating the effect of heat on proteins. Eggs contain all kinds of different proteins—an egg white alone has more than a hundred different types. All of these proteins denature and coagulate at slightly different temperatures. What's so remarkable is that we can observe on a macroscopic scale the step-by-step molecular transformations in an egg as we add heat. In fact, the transitions are so predictable that a trained chef can use eggs almost as thermometers. Let's look at how this works.

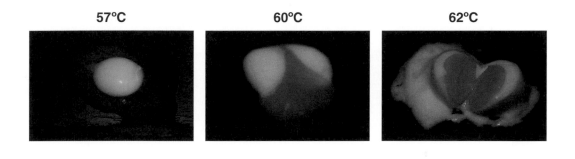

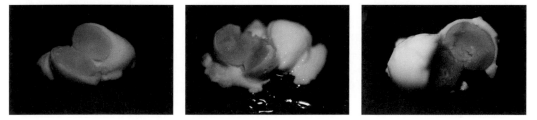

FIGURE 6 Sous vide cooking is a technique used by chefs who have a very specific temperature in mind when they are cooking a dish. It usually involves vacuum-sealing the ingredients to keep them clean and putting the package into a water bath held at the target temperature. After several hours (depending on the size of the pieces), the food temperature will be the same as the water bath temperature. This technique is useful in commercial kitchens because different ingredients can be held at a temperature, ready for finishing and plating. Sous vide can be used to create remarkable dishes that require stringent temperature control. The figure shows eggs cooked at different temperatures between 57°C and 69°C (134.5°F and 156°F). In the text we describe these remarkable transitions.

Making sous vide eggs doesn't need to involve fancy, expensive equipment. By understanding the principles of the technique, you can easily make them at home. The following recipe is an exercise from our online HarvardX class. Participants around the world who took the class perfected the procedure in various ingenious ways, and the images above are examples of some of them.

Sous Vide Eggs

Ingredients

12 eggs, at room temperature

Directions

1. Place 3 eggs in a medium pot and cover with about 1 inch of water. Bring the water to a boil. Immediately turn off the heat and let the eggs cook for 10 to 15 minutes. Remove the eggs with tongs or a slotted spoon and set aside to cool; do not discard the water in the pot. These hard-boiled eggs will serve as a comparison to the other eggs.

2. Fill another pot (or kettle) with water and bring to a boil.

3. Measure the temperature of the water left over from boiling the eggs. If it is greater than 62°C (143.5°F), let it cool until it reaches this point. If it is lower than 62°C (143.5°F), add boiling water until it reaches this temperature.

4. Place 3 uncooked eggs in this 62°C (143.5°F) bath. Keep the bath between 58°C (136°F) and 62°C (143.5°F) for at least 20 minutes (or ideally, up to 40 minutes) by adding boiling water to the bath as it loses heat to the eggs and the surroundings. The temperature range doesn't need to be exact, but the bath should not sit above 62°C (143.5°F) for more than a few seconds at any given point. The target temperature for the final eggs will be the average of the upper and lower limits (60°C/140°F). It might be helpful to note how long it takes for the water bath to drop 1 degree Celsius so that you don't have to monitor the temperature of the bath constantly (the time will vary based on a number of factors, but a good starting estimate would be that 1 degree Celsius is lost every 1.5 minutes).

5. At the end of the elapsed time, remove one egg from the bath, cool it briefly under cold running water, and then crack the egg into a bowl. If it looks to be the appropriate consistency as described in the text, remove the other 2 eggs and set them aside. If not, leave the other 2 eggs in the bath for another 10 and 20 minutes, respectively, then run under cold water and crack into separate bowls.

6. Repeat steps 3 and 4 using 3 more eggs in a water bath in the range of 61°C–65°C (142°F–149°F) to obtain eggs at a final temperature of 63°C (145.5°F), and then the remaining 3 eggs in a water bath in the range of 64°C–68°C (147°F–154°F) to obtain eggs at a final temperature of 66°C (151°F). Compare to the hard-boiled eggs from step 1. For each of the temperature ranges, estimate the internal temperature of the egg by comparing the doneness of the egg to the consistencies described in the text. ❀

We have already mentioned that an egg cooked in a 57°C (134.5°F) water bath for a couple of hours will no longer contain dangerous microorganisms, but it still has a liquid yolk and is transparent white. If you do the same in a water bath that is only 3 degrees Celsius warmer (60°C/140°F), the egg white will just barely set and change color. Some of the proteins in the egg will have started to denature and coagulate, but not all of them. At 62°C (143.5°F), the egg white has changed color and holds together, but the yolk is still runny. (This is perfect for eggs Benedict.) At 64°C (147°F), the yolk has fully set but is soft and custardy. So within only 2 degrees Celsius, from 62°C to 64°C (143.5°F to 147°F), the yolk has gone from being completely liquid to completely set. The transition is so sudden that someone familiar with these transitions can tell the temperature to within a couple tenths of a degree. Is the water bath 63°C (145.5°F) or more like 63.3°C (146°F)? An egg will tell you.

A 64°C (147°F) egg is often considered the "perfect egg" and is used in restaurants all the time. Chefs use the controlled temperature technique described in the sidebar to produce large quantities of these "perfect" eggs that can easily be served to customers as they are ordered. But any egg can be considered "perfect" depending on your cooking plans. Only 1 degree Celsius higher, at 65°C (149°F), the yolk changes further and acquires the texture of Play-Doh. At 66°C (151°F), the egg yolk turns into a slightly firmer marzipan-like texture, which is ideal for molding into different shapes and rolling into thin sheets, which can't be done without the yolk cracking at higher temperatures. At 67°C (152.5°F), the yolk loses its creaminess and starts to acquire some of the granular texture of a traditional cooked egg yolk. Finally, at 70°C (158°F), the egg starts to look more like a typical soft-boiled egg, except the yolk has the same consistency throughout instead of having the center be slightly more cooked than the outside. It isn't until around 75°C–80°C (167°F–176°F) that the egg becomes "hard-boiled," including having the characteristic greenish color and sulfury smell.

Each macroscopic transformation corresponds to a different molecular transformation. At 63°C (145.5°F), an egg white protein named ovotransferrin unfolds. As you increase the temperature, other yolk proteins unfold—ovalbumin, ovomucoid, ovoglobulin, and many more. What's so remarkable is that all of these transitions occur within the narrow span of 13 degrees Celsius. Perhaps nowhere else in cooking is accurate temperature control as important as here. Even a degree off can result in an incorrectly cooked egg.

It is important to note that an egg left at 64°C (147°F) won't stay as a "perfect" egg forever. If kept at this temperature, the proteins will slowly continue denaturing and change the texture. But within typical cooking time frames, this technique is an incredibly accurate and repeatable way to cook eggs.

PROTEIN UNFOLDING AND GELATION

Let's take a closer look, at the molecular level, to see what is going on with these protein transitions: the twenty different amino acids that make up proteins are arranged into unique sequences, and the exact sequence is critical to how the protein works. This is at the heart of biology, and many of the control circuits in biology are designed to keep the proteins functioning in different conditions. But that is not our challenge when cooking: when we cook, we create delicious food by modifying the native proteins so that they can no longer fulfill their original biological function.

The twenty amino acids have different chemical and physical properties, but two properties are especially important for what happens to proteins when heated: hydrophilicity and hydrophobicity. Hydrophilic amino acids dissolve in water;

the word comes from a Greek word that means "water loving." Hydrophobic, or "water fearing," amino acids, on the other hand, don't dissolve in water, but dissolve in oil instead. Just like oil and water, hydrophilic and hydrophobic amino acids do not like being in close contact with each other, but prefer sticking to others of their own type.

The hydrophobic amino acids on a protein chain try to assemble together and stay away from the water surrounding the protein as much as possible. The hydrophilic amino acids also assemble together, but they do the opposite and stay in contact with the water. Thus, when a protein folds into its preferred structure, something like the following scenario will take place: If the temperature is low enough so that the protein doesn't wiggle a lot, it will try to fold into a compact structure. It will try to make all of the hydrophobic amino acids squish together in the interior of the protein so that they don't contact the water. The hydrophilic amino acids, by contrast, will assemble on the surface, in contact with water and away from the hydrophobic interior. It's not possible to do this arrangement perfectly because the amino acids are also attached to each other in the protein string, but the idea is to minimize exceptions to this rule as much as possible. This follows the principle of minimizing the free energy of the protein, and the configuration that is the most favorable will win.

At high temperatures, the protein wiggles from the heat, and this causes the protein to unfold quite a bit. But the hydrophobic residues will still try to stick together, as will the hydrophilic residues. So, in a somewhat random fashion, the newly exposed hydrophobic residues will stick to other hydrophobic residues, and hydrophilic residues will stick to other hydrophilic residues. The residues on one protein form links with those on other proteins, and a network of crosslinks begins to emerge. This network then completely transforms the appearance and the properties of the proteins.

In the specific example of cooking an egg, the process goes as follows: Egg white primarily consists of water, and suspended in the water are several different types of proteins; one of the main ones is called ovotransferrin. When the egg is raw, the ovotransferrin proteins are perfectly folded into their native structure and look like little blobs suspended in the water. When heated to about 63°C (145.5°F), the ovotransferrin proteins start to unfold and the polymers spread out in the water and do something totally remarkable.

To explain this, we need you to pause for a moment and use your imagination. Consider a group of kids running around in a kindergarten. You are the teacher, and your job is to convince them to stop moving. To do this, you tell them loudly to stop. Some of them don't listen and still move. You then tell them to hold hands with two of their neighbors. The kids like each other, so this makes them happy. At that point, everyone is stuck. None of the children can move without letting go. A similar thing happens with proteins in an egg. They unfold and then stick to each other because the exposed hydrophobic amino acids are sticky. At that point, none of the proteins can move anymore. The material thus transforms from a liquid to a solid. The liquid egg has completely changed. Remarkably, though, the egg is still primarily water, and it is only the small amount of coagulated protein that spreads through the whole egg that causes it to be solid, and hence cooked. In addition, the coagulated proteins also scatter light much more, causing the color to change and the egg to turn white.

This transition in an egg, when it goes from liquid to solid, is called a *gelation* transition. Something similar happens when we heat other foods that are rich in proteins, such as steak and fish. In a raw steak, the proteins are folded and the steak is tender and appears red and slightly translucent. With heat, usually between 50°C (122°F) and 60°C (140°F), the meat's primary proteins, myosin and myoglobin, start to unfold and form crosslinks. The more crosslinks, the firmer the meat becomes, as it goes from rare to medium to well-done. So just as in the egg, the microscopic changes that occur as you cook a steak are directly related to both the color and the mouthfeel of the final steak.

Cooking Pasta

To complete our tour of the effect of heat on the different food components, let's briefly revisit carbohydrates. We've already seen what happens when we heat

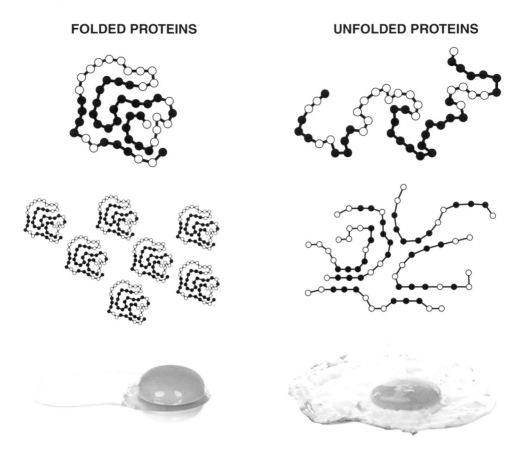

FOLDED PROTEINS

UNFOLDED PROTEINS

FIGURE 7 A schematic of an egg protein in its folded and unfolded state. The black dots represent hydrophobic amino acids, which do not want to interact with water. The white dots represent hydrophilic amino acids, which like to be in the presence of water. In the folded state, most of the black dots are at the center of the blob where there is less water. These tightly folded proteins remain separate in the egg white, which is mostly water—this is why the white is liquid when it is uncooked. After cooking, though, the proteins unfold so that the hydrophobic residues are now exposed. Once unfolded, the hydrophobic residues are still hydrophobic, so they find similar regions in other proteins to interact with. This results in coagulation, where the unfolded proteins are now stuck to each other. The cooked egg whites are white because the coagulated proteins scatter light.

simple carbohydrates such as sugar. But what about complex carbohydrates? Pasta contains a huge amount of complex carbohydrates, namely starches. A starch is long chains of sugar molecules attached to each other. What happens when we put it in boiling water? First, since pasta is dehydrated, it gets rehydrated when submerged in water. The effect is that the starch swells. Second, applying heat helps both the rehydration and swelling, causing the starch to leak proteins, which causes the starch granules to stick to one another. The pasta was previously dry and brittle. After applying heat and water, it has now turned into a gel, opaque and squishy, and not unlike the cooked egg white.

This is the effect of heat on food, all summarized in the image below. But heat is only one of many ways to transform food. Next we will look at other transformations.

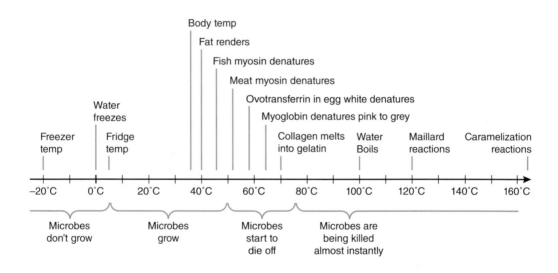

CHAPTER 3

Charge, pH, and Enzymes

Enzymes and acids are hidden ingredients in recipes. They are critical to how dishes taste and also how they cook. Without enzymes or acids, we couldn't even digest food, let alone eat it, so it is not surprising they play such a central role.

—Harold McGee

Transformations with Electric Charge

In the last chapter, we discussed how temperature transforms food in the cooking process. There is another control knob that we can use to cook, but it is a bit less familiar than temperature. This knob is electric charge. You may remember learning about electric charge in your high school physics class. There you were taught that like charges repel and opposite charges attract. This is true, but you have to admit, it's a bit boring. Well, what do charges have to do with cooking? In fact, not only are charges associated with cooking, they also play an important role in many of our most loved foods, as we will soon see.

As it turns out, both proteins and carbohydrates, two of the main food molecules, can have electric charge. Proteins are charged because a few of their amino acids are charged (aspartic acid and glutamic acid have negative charge, and arginine and lysine have positive charge). This means that when you add other substances to food that is already charged, the proteins and carbohydrates that make up the food are disrupted, thus changing the texture of the food as a whole. Strikingly, this change in texture is often similar to what can otherwise occur with heat.

TRANSFORMATIONS WITH PH

The most common ingredients that are added to food to manipulate charge are those that change the pH, making the food either more acidic or more basic. In fact, pH is a particularly dramatic way to affect charge, as it actually changes the charge on the proteins and carbohydrates. Let's illustrate this with a few examples.

THOUSAND-YEAR-OLD EGGS

The sidebar shows the recipe for a thousand-year-old egg. The ingredient list includes a combination of things you would normally not expect to see in recipes: salt, lye, and tea leaves! In traditional recipes for century eggs, the ingredient list can be even more esoteric and include materials like clay and ash. What all of these recipes have in common is that they have an alkaline pH. All of them also include an important ingredient: time. The eggs must be immersed in the solution for a really long time, sometimes up to six weeks as in Corey Lee's recipe, featured in the sidebar. Imagine waiting six weeks for eggs! But this is what is required when cooking without heat. Just as when heated, the egg becomes a solid that cannot be changed back into a liquid. As with a heated egg, the color also changes, but it does so even more dramatically: the century egg white turns translucent auburn, and the yolk takes on a variety of blue-green-yellow shades.

Century Egg

Ingredients

Pickled Ginger (recipe follows)
Potage (recipe follows)
Cabbage Juice (recipe follows)
1 Thousand-Year-Old Quail Egg half (recipe follows)
Salt

Directions

1. Place a dollop of chopped ginger in a small bowl.

2. Add a large spoonful of heated potage on top.

3. Cover the surface with the cabbage juice.

CHARGE, pH, AND ENZYMES

4. Season the egg half with ginger pickling juice and salt. Place on top.

Pickled Ginger

—————

Ingredients

1 (2-inch) piece fresh ginger, peeled and sliced very thin on a mandoline

2 parts water

1 part champagne vinegar

1 part sugar

Directions

1. Weigh the sliced ginger and put it in a large heatproof bowl.

2. Measure the appropriate amounts of water, vinegar, and sugar into a large pot and bring to a boil over high heat.

3. Pour the boiling pickle liquid over the ginger. Let it come to room temperature.

4. Cover and store in the refrigerator for at least 1 week.

5. Strain and finely chop the ginger, reserving the juice.

Potage

—————

Ingredients

200 g bacon, sliced thin

700 g green cabbage, sliced thin

200 g onion, sliced thin

50 g unsalted butter

1 L chicken stock

7 g salt

0.5 g cayenne pepper

100 g heavy cream, heated

Directions

1. Melt the butter in a large pot over medium heat. Add the bacon, cabbage, and onion and cook until softened but not caramelized.

2. Add the chicken stock, turn up the heat to medium-high, and bring to a boil. Reduce the mixture to 900 g.

3. Remove from the heat, add the cream, salt, and cayenne, and purée with an immersion blender or food processor.

4. Pass the mixture through a fine-mesh strainer into a bowl set over an ice bath. Set aside until completely chilled.

Cabbage Juice

Ingredients
½ head green cabbage

Per the weight of the juice:
2% Ultra-Tex starch
1% salt
0.5% champagne vinegar

Directions
1. Bring a large pot of water to a boil over high heat. Fill a large bowl with cold water and ice cubes.

2. Peel the outer green leaves of the cabbage. Thinly slice a few slivers of the remaining cabbage and set aside.

3. Blanch the cabbage for 1 minute in the boiling water, then transfer to the ice bath to shock them.

4. Squeeze out the excess water and pat dry.

5. Pass the cabbage leaves through a juicer.

6. Take all the pulp and run it through the juicer one more time.

7. Pass the juice through a chinois.

8. Weigh the juice and add the appropriate amounts of starch, salt, and vinegar.

9. Add a few slivers of the sliced cabbage to each serving.

Thousand-Year-Old Quail Eggs

Ingredients

330 g water

1 g pu-erh tea

16 g kosher salt

14 g food-grade sodium hydroxide (lye)

0.7 g food-grade zinc

24 quail eggs

Directions

1. Bring half of the water to a boil in a small saucepan.

2. Remove the pot from the heat and add the tea. Steep for 20 minutes.

3. Add the salt, lye, and zinc and stir to dissolve. Add the rest of the water.

4. Transfer the mixture to a ceramic jar with a lid and set aside overnight.

5. The next day, submerge the eggs in the brine. Cover the jar and keep the eggs in the brine for 12 days. (Note: Diffusion takes a long time! Chapter 4 will explain why.)

6. Remove the eggs from the brine and dip them briefly in water to rinse the shell, then let the eggs air-dry for about an hour.

7. Seal the eggs in a ziplock plastic bag. Place the sealed plastic bag in a container that doesn't allow light to enter. Store in a cool, dry place for at least 4 weeks.

8. Bring a large pot of water to a boil. Fill a large bowl with cold water and ice cubes.

9. Remove the eggs from the bag and drop them into the boiling water for 1 minute. Boil only a few eggs at a time so the water does not stop boiling.

10. Use a slotted spoon to transfer the eggs to the ice bath to shock them. Leave them in the ice bath until they have completely cooled.

11. Peel the eggs, then cut them in half lengthwise.

Century eggs are a preserved egg delicacy in China, typically made with duck, chicken, or quail eggs. The whites have a jelly-like texture and the yolk is creamy but has a slightly pungent eggy smell. If you have only ever had eggs cooked with heat before, the taste of a century egg will be a new experience, yet it still has some of the familiarity of an egg. Coating raw eggs with a clay made of tea, various salts, and wood ash provides the alkaline environment necessary for the chemical reactions to take place. As with any brined or pickled food, it takes a while for the process to finish—over a month in this case. If the conditions are right, some of the salt may even crystallize at the surface of the whites, creating beautiful patterns similar to snowflakes (the Chinese call these "pine flower" century eggs). ✳

LUTEFISK

Another recipe that uses pH as the key ingredient is the Nordic delicacy lutefisk. One of the authors of this book, Pia, grew up in Sweden and recalls eating lutefisk on Christmas every year as a child. (Not being a fan of the fish itself—you'll see why if you make this recipe—she mostly ate the accompanying potatoes, peas, and mustard sauce.) Lutefisk originates with an age-old method of preserving fish. The freshly caught fish is air-dried for many weeks until the water has evaporated and the only thing left is the dry, hard flesh. To make it edible, there is a lengthy procedure of alternating soaking the flesh in lye and water. This goes on for close to a month—traditionally the designated start date was Anna Day, December 9, leaving enough days for the chemical and physical transformations to occur in time for Christmas. (Of course, even in Sweden, these days you can just go buy the finished product in the supermarket.) During the long cooking time, the basic lye solution slowly breaks down the protein chains. As a result, the fish acquires a gelatinous texture. It's a little bit like eating fish jelly.

Lutefisk

Ingredients
1 piece stockfish (dried cod or ling)
2 tablespoons food-grade sodium hydroxide (lye)
Salt

Directions
1. Place the dried fish in a container of water and refrigerate for 6 days, changing the water daily. Drain.

2. Wearing gloves, dissolve the lye in 4 liters of fresh water by slowly adding small amounts of the lye to the water at a time. Do NOT do the opposite and add water to the lye as it can cause a small explosion out of the container. Add the fish and refrigerate for 3 days.

3. Wearing gloves, remove the fish. Transfer to a container of fresh water and refrigerate for 4 to 6 days, changing the water daily.

4. Test if the fish is ready by boiling small pieces; if the consistency is too gelatinous, soak for longer; if it is too hard, start over. The fish should be flaky and have a mild taste.

5. To cook, rinse the fish and place it in a stainless steel pan (aluminum and copper pots will react with the lye).

6. Add salt to taste, cover, and cook over low heat for 20 minutes.

7. Remove the fish and discard the liquid.

8. Serve with boiled potatoes, peas, and melted butter or a mustard cream sauce.

Lutefisk is made by rehydrating dried fish in an alkaline solution of lye (sodium hydroxide) or slaked lime (calcium hydroxide). Both solutions are highly alkaline. You can calculate just how alkaline by studying the recipe above in more detail. Sodium hydroxide (NaOH) has a molecular weight of 40 g/mol. One tablespoon of sodium hydroxide weighs about

32 grams, so this means we're adding 1.6 moles of sodium hydroxide to the 4 liters of water, making the concentration 0.42 moles per liter. This corresponds to a pH of 13.6! This is such a high pH that you have to wear gloves to avoid getting burned. The many steps of soaking and rinsing removes the lye, but even so it's recommended you discard the liquid after cooking. Using alkaline solutions to prepare food is common in many cultures. Native American cuisine, for example, along with both Mexican and Central American cooking, use lye and slaked lime to prepare dried corn, which is then used in a number of preparations. ⚛

CEVICHE

Both century eggs and lutefisk are cooked in alkaline pH, be it in the form of lye, clay, or ash. There are, however, dishes that do the same thing using acids. One of the best known and loved is ceviche. Ceviche is raw fish that has been immersed in lime or lemon juice for varying amounts of time. The low pH of the citrus juice cooks the fish, and the effect is readily observable with the naked eye. A piece of white fish such as tilapia will go from translucent white to opaque. A piece of red tuna will go from brilliant red to dull pink. The fish also acquires a texture that is slightly tougher, not unlike the texture of fish that has been cooked with heat. The sidebar shows a classical ceviche recipe and a Philippine version, kinilaw, which uses vinegar instead of lime juice.

SIDEBAR 3: CEVICHE

Ceviche

Ingredients
200 g sashimi-grade fish, cut into 0.5 cm pieces
250 mL lime juice

1 tomato, diced

1 or 2 garlic cloves, finely minced

½ red onion, finely diced

2 tablespoons chopped fresh cilantro

1 serrano pepper, seeded and finely chopped (optional)

1 teaspoon salt

¼ teaspoon ground black pepper

1 avocado, pitted, peeled, and sliced, for serving

Tortilla chips, for serving

Directions

1. Pour the lime juice into a medium bowl. Add the fish and marinate at room temperature or in the refrigerator for 15 to 20 minutes.

2. While the fish is marinating, toss together the tomato, garlic, onion, cilantro, serrano (if using), salt, and pepper in a small bowl.

3. Drain the lime juice from the bowl and mix the fish with the vegetables. You can cut the pieces in half and examine how the color and texture changed due to the lime juice.

4. Serve portions on individual plates and garnish with slices of avocado. Enjoy with tortilla chips. ⚛

SIDEBAR 4: MARGARITA FORÉS'S KINILAW

Kinilaw

Ingredients

Brine/Sauce

¾ cup white vinegar

¼ cup chopped calamansi

1 medium red onion, chopped

1 (1-inch) knob ginger, peeled and chopped

3½ tablespoons white sugar

½ tablespoon salt

Kinilaw

¼ cucumber, thinly sliced

2 small red radishes, thinly sliced

2 green finger chiles, thinly sliced

½ small yellow bell pepper, seeded and diced

30 g Boy Bawang (Filipino toasted corn snack; can substitute boiled corn)

250 g sashimi-grade tuna, sliced or chopped

Adlai crackers or other light crackers, to garnish

Microgreens, to garnish

Directions

1. In a bowl, combine all the ingredients for the brine and stir well. Measure out ¼ cup into another bowl and refrigerate the rest.

2. Add the cucumber to the bowl with the reserved ¼ cup brine. Refrigerate until pickled (at least 30 minutes, or overnight), then drain.

3. In another bowl, toss together the radishes, chiles, bell pepper, Boy Bawang, drained pickled cucumber, and tuna. Pour in the remaining brine and toss. Transfer to a plate.

4. Top with the adlai cracker and microgreens. ⚛

The pH Dependence of Protein Unfolding

From the examples just described, it's clear that pH can have a profound effect on the visual and textural qualities of foods. What happens on the microscopic scale that explains these changes?

Recall our folded protein chain with all of its different amino acids having distinct properties—some are like water, some are like oil, and a small number of them have positive or negative charge. If two positively charged amino acids are close to each other, they repel one another and slightly push that part of the protein chain apart. Two amino acids of opposite charge will instead attract and pull their chains together. When the protein is perfectly folded, all of these attractions and repulsions will be balanced, and the protein will be fully stable.

We saw what happens when you add heat to such a protein: it starts to wiggle from the extra energy causing the bonds to break. The protein becomes denatured and then coagulates as its various parts form new bonds to other random parts of the protein chain. This is similar to what happens with electric charge. In a solution with low pH, the huge number of positively charged hydrogen ions will bind to the amino acids wherever they can. Everywhere they bind, they neutralize the negative charges and add positive charge to the already neutral ones. This throws off the balanced structure. The chemical properties that had so carefully been accounted for are now completely changed, since there are now more positive charges everywhere. They push the various protein parts apart, and the protein denatures in a similar way to what happened with heat. Once the proteins are denatured and unfolded, they can interact with other proteins to coagulate, in the same way they do with heat.

When you cook with basic pH, a similar phenomenon happens, except now the charges are skewed in the opposite direction. There is a lack of positive hydrogen ions, so the protein will be stripped of its positive charges and become too negative overall. This similarly destabilizes the protein, and it falls apart.

Looking closely at the microscopic level explains why cooking with heat versus pH can have somewhat similar effects on food—on a microscopic scale, the changes are actually not all that different.

The likelihood of a protein denaturing depends on how much charge you add and the particular properties of the protein being cooked. The sidebars shows the pH dependence of protein unfolding for two common proteins: albumin, one of the main proteins in egg white, which gets "cooked" in the century egg recipe, and fish myosin, the protein that is denatured when cooking ceviche and lutefisk.

The graphs in Figure 2 show what happens to the net charge of the protein when you change the pH. For a neutral pH, at which most proteins are balanced, the net charge is usually close to zero. Often you have to change the pH quite a

A

Amino acid	Charge at pH 7	Amino acid	Charge at pH 7
Aspartic Acid (D)	-	Alanine (A)	hydrophobic
Glutamic acid (E)	-	Glycine (G)	hydrophobic
Arginine (R)	+	Valine (V)	hydrophobic
Lysine (K)	+	Leucine (L)	hydrophobic
Histidine (H)	hydrophilic	Isoleucine (I)	hydrophobic
Asparagine (N)	hydrophilic	Proline (P)	hydrophobic
Glutamine (Q)	hydrophilic	Phenylalanine (F)	hydrophobic
Serine (S)	hydrophilic	Methionine (M)	hydrophobic
Threonine (T)	hydrophilic	Tryptophan (W)	hydrophobic
Tyrosine (Y)	hydrophilic	Cysteine (C)	hydrophobic

B

Folded   Unfolded

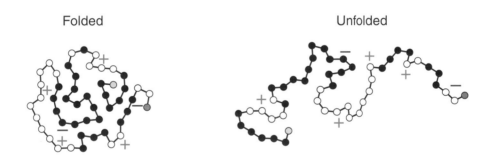

FIGURE 1 **(A)** All proteins are pieced together from a set of twenty different amino acids, listed in this table with the single-letter abbreviations for each amino acid within parenthesis. Of the twenty amino acids, two are positively charged and two are negatively charged. The rest are uncharged but can have either hydrophilic or hydrophobic properties.

    (B) A naturally folded protein (left) tends to have hydrophobic amino acids in the center, away from water (black dots), and charged and hydrophilic amino acids on the surface (white dots). A change in pH is the equivalent of adding lots of positive or negative charges in the form of additional H^+ or OH^- ions. For example, if you lower the pH, H^+ ions will bind in various places on the protein and make it overall more positively charged. Since all the positive charges repel each other, this destabilizes the protein and it will partly unfold. The newly exposed amino acids will then interact with other amino acids on the same or other proteins. To the naked eye, the color and texture of the food will change as its proteins undergo these changes.

bit before the net charge changes, as the structure of the protein has evolved to withstand some fluctuation in pH without falling apart. But at some point, when the pH is either too low or too high, the protein can't absorb any more charges and the net charge changes dramatically. At this point you also can expect to see a change in the color and texture on a macroscopic scale.

For example, let's see what these graphs can tell us about ceviche. Look at the plot for fish myosin. In its natural state (at neutral pH 7), the fish myosin has an

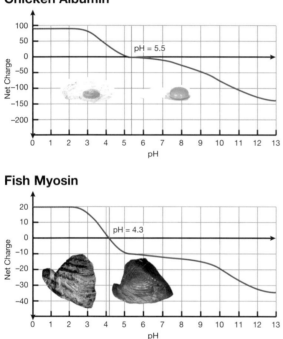

Chicken Albumin

Fish Myosin

Chicken ovalbumin (major component in egg white)
386 Amino Acids

MGSIGAASMEFCFDVFKELKVHHAN
ENIFYCPIAMSALAMVYLGAKDSTRT
QINKVVRFDKLPGFGDSIEAQCGTSV
NVHSSLRDILNQITKPNDVYSFSLASRL
YAEERYPILPEYLQCVKELYRGGLEPIN
FQTAADQARELINSWVESQTNGIIRN
VLQPSSVDSQTAMVLVNAIVFKGLW
EKAFKDEDTQAMPFRVTEQESKPVQ
MMYQIGLFRVASMASEKMKILELPFA
SGTMSMLVLLPDEVSGLEQLESIINFE
KLTEWTSSNVMEERKIKVYLPRMKM
EEKYNLTSVLMAMGITDVFSSSANLS
GISSAESLKISQAVHAAHAEINEAGRE
VVGSAEAGVDAASVSEEFRADHPFLF
CIKHIATNAVLFFGRCVSP

FIGURE 2 **Left:** The graph at the top left shows the pH dependence of chicken serum albumin, one of the main proteins in eggs. The graph at the bottom left shows the pH dependence of fish myosin protein, the main protein in fish. Both graphs show the total charge on the protein at each pH. As the pH increases, both proteins become more and more negatively charged. When the net charge changes quickly over a short range of pH values, as it does at around pH 4 for both the egg and fish proteins, there is likely a clear transition in the proteins.

Right: The amino acids in proteins are usually shown as a long string of letters, where each letter corresponds to an amino acid. Here we see chicken ovalbumin, which consists of 386 amino acids. In the image, amino acids with negative charge, D and E (aspartic acid and glutamic acid), are colored red, and amino acids with positive charge, R and K (arginine and lysine), are colored green. Although only a small number of the amino acids have a charge, they play an important role in the folding and functioning of the protein.

approximately neutral charge. Here it is stable and folded into its functional structure. But at pH 2, the charge is much fold higher—we can expect it to be very unstable and denatured. In between these pHs, at pH 4, the net charge is somewhere in the middle. These data points do indeed correlate to real life: lemon juice (pH 2) can "cook" ceviche, but orange juice (pH 3.5–4) cannot; it is not sufficiently acidic to affect the charge. Similarly, the lye solution in the lutefisk recipe has a pH of ~13, which is very basic. At pH 13, fish myosin has a net charge of –27 according to our diagram. So you can expect the fish to look cooked at this pH, and this is indeed what happens. With lutefisk, not only does the protein denature, but it also starts to break down, which contributes to the gelatinous texture. We will discuss the scientific reasons for this when we discuss elasticity in chapter 5.

Transformations with pH and Heat

RICOTTA CHEESE

Authentic Italian ricotta cheese is typically made from the whey left over after making other cheeses—*ricotta* means "recooked." But these days, it is often made from milk instead. Contrary to what most people think, ricotta is incredibly easy to make. You simply heat milk, add some vinegar or lemon juice, wait for the milk to curdle, strain off the liquid, and—voilà!—ricotta cheese!

You may be thinking, did they just say add *vinegar* to cheese? While you may be more used to cooking with unconventional ingredients when making ceviche or lutefisk—these are, after all, foods with very special tastes and textures—it seems much more out of place in a food like cheese. In addition, vinegar on its own isn't very delicious, and a cheese that tastes like vinegar certainly doesn't sound very appetizing.

You could, in theory, make ricotta with vinegar alone and not add heat. The vinegar would indeed curdle the milk, since its pH is low enough to significantly affect the net charge. But the taste would be terrible. In theory, you could also curdle the milk by heating alone and not add vinegar. But in practice, you wouldn't be able to add enough heat to the milk for this to happen—milk is mostly water, so the temperature wouldn't go higher than 100°C (212°F). Instead, you must very delicately balance the two: add just enough vinegar so that the milk curdles with some appropriate amount of heat, but not enough to significantly alter the taste.

On a microscopic scale, the many charges from the vinegar flood the casein and whey proteins in the milk and destabilize them. The proteins are destabilized enough so that even just a small amount of heat is enough for them to fall apart completely.

Achieving this balance is not trivial. In fact, while making any ricotta cheese is relatively easy, making perfect ricotta cheese can be incredibly difficult. The recipe shows how you can experiment with this in your kitchen. Heat it too much and the proteins become overcooked, so you get grainy, rubbery curds. Heat it too little and the milk does not curdle enough. Add too much vinegar and the ricotta will taste sour. Add too little and you get no curds at all. The perfect ricotta recipe is a careful balance of all of these factors, resulting in soft, creamy curds.

SIDEBAR 5: RICOTTA

Ricotta

Ingredients
8 cups (2 L) whole milk
1 cup (240 mL) heavy cream
½ teaspoon (2.5 mL) salt
¼ cup (60 mL) lemon juice or white wine vinegar

Directions
1. In a large saucepan, heat the milk, cream, and salt over low heat until the mixture reaches 200°F (93°C), stirring regularly to prevent the bottom from burning. Remove the pan from the heat.

2. Add the lemon juice or vinegar and quickly stir for a few seconds to combine. Stop stirring and let the mixture rest for 3 to 5 minutes to allow the curds to coagulate. Do not disturb the pan during this step. The longer you leave the mixture, the larger the curds will be.

3. Line a strainer with a double layer of cheesecloth, making sure to leave plenty of extra cheesecloth hanging over the edges. Place the strainer in the sink or a larger bowl and gently pour the ricotta mixture through the cheesecloth. Let it drain for however long you desire—5 minutes will result in a soft, spreadable, ricotta, and 1 hour will result in a much firmer ricotta.

4. Empty the cheese into a serving bowl and garnish as desired (salt, pepper, honey, rosemary, etc.). (The liquid left in the bowl, called whey, can be discarded or used in smoothies or other recipes.) ❀

POACHED EGGS

Managing the delicate balance between heat and pH can solve tricky cooking dilemmas, such as those posed by poached eggs. Poached eggs are prepared by simmering peeled eggs in water for short periods of time. The result is delicious: the white is softly cooked, the yolk smooth and gooey. Successfully poached eggs are, however, notoriously difficult to make. The egg should remain an intact blob, but how do you crack a raw egg into a pot of boiling water without having it immediately disperse into the water and cook into a scrambled egg mess? Somehow you must find a way to make the uncooked egg hold together for long enough that the surface solidifies from the heat and holds everything together while the interior cooks.

Cookbooks offer many solutions to this conundrum. One is to briefly swirl the boiling water with a spoon to create a vortex that holds the liquid egg together. Another is the clever use of a low-pH ingredient. This is the role of the vinegar called for in the recipe in the sidebar. It makes the surface of the raw egg cook by both acid and heat. Thanks to this double cooking approach, a cooked outer layer forms relatively quickly—much faster than would be the case with heat alone—and this solid casing helps hold the egg together so it does not disperse in the pot. By adding just a small amount of vinegar, small enough so that you cannot taste it, you effectively increase your chances of making perfectly poached eggs.

SIDEBAR 6: POACHED EGGS

Poached Eggs

Ingredients

8 cups (2 L) water
2 teaspoons (10 mL) vinegar
4 large eggs

Directions

1. Combine the water and vinegar in a wide, heavy saucepan and bring to a simmer over medium heat.

2. Break the eggs into individual ramekins or small bowls.

3. Using a slotted spoon, create a slow whirlpool in the center of the pan. Gently slide an egg into the center of the whirlpool and allow it to poach at a simmer until the white is firm but the yolk is still runny, 2 to 3 minutes.

4. Using the slotted spoon, transfer the egg to a paper towel to dry. Repeat with the remaining eggs, one by one. ⚛

Transformations with Salt: Curing and Brining

We saw what happens when we cook with pH—the effects were due to either the abundance or absence of charged hydrogen ions. But what about cooking with other charges? Table salt, for example, is made up of one positive ion and one negative ion. Could we cook with it? The century egg recipe already gives a hint: we added salt in addition to the lye and tea leaves. In fact, if you don't have access to lye, you can cook a version of a century egg using only salt. Salt alone has a similar effect to pH because it also changes the net charge of the proteins. But in addition to affecting the proteins alone, salt can affect cells as a whole, and this property can be highly useful when cooking. The recipes in the following sidebars show two examples: coleslaw and jerky.

Coleslaw

Ingredients
500 g (5 cups) shredded purple and green cabbage
2 teaspoons (10 mL) salt
50 g (1 cup) shredded carrots

Directions
1. In a large bowl, toss the cabbage and carrots with the salt. Set aside for 10 to 15 minutes.

2. Drain off the liquid.

3. Add your dressing of choice (mayonnaise or apple cider dressing both work well) and serve.

When you set aside the cabbage and salt, the cabbage starts to leak water. If you tip the bowl after 15 minutes or so, you'll see a good amount of liquid at the bottom. What's happened here is that the water on the outside of the cabbage cells is much saltier than the water on the inside of the cells. There is a natural process called osmosis that follows from these types of concentration differences, which is that they strive to become more equal. So the water leaks out of the cells in an effort to dilute the outside concentration. As you see, this is a case in which the entire cell is affected, not just the proteins. The result is that the cabbage feels crunchier. By draining the water early on in the recipe, you prevent the cabbage from leaking water later on, when the coleslaw is finished. So this technique makes for crunchy, non-watery coleslaw. ❀

Carrot Jerky

Ingredients

4 liters (4.3 quarts) purified water, at room temperature

100 g (1 cup) calcium hydroxide

20 young carrots, scrubbed and greens trimmed to 1.25 cm (½ inch)

1475 g (7⅔ cups) granulated sugar

475 g (2⅔ cups) packed brown sugar

485 g (2 cups) fine sea salt

210 g (7.3 ounces) fresh ginger, peeled

1 head garlic, cloves peeled and roughly smashed

10 g (4⅓ teaspoons) cracked black pepper

450 g (1 pound) large carrots

Cayenne pepper

400 g (14 ounces) orange slices (from 2 to 3 oranges)

10 whole black peppercorns

32 g (1.1 ounces) Pure-Cote

300 g (¾ cup) liquid glucose

Directions

At least 4½ days before serving:

1. To make the pickling lime, mix the purified water with the calcium hydroxide in a large tub or pot.

2. Soak the young carrots in the pickling lime for 3 hours.

3. Meanwhile, to make the dry cure, combine 575 g (3 cups) of the granulated sugar with the brown sugar and 450 g (1¾ cups plus 2 tablespoons) of the fine sea salt in a large ziplock bag. Dice 10 g (0.35 ounce) of the peeled ginger. Add the diced ginger, garlic, and cracked black pepper to the bag and shake well.

4. When the young carrots are finished soaking, rinse them in cool water and pat dry with paper towels. Add the young carrots to the bag containing the dry cure and jostle the bag to ensure that they are thoroughly coated. Seal the bag and allow the young carrots to cure at room temperature for 72 hours. (Note that water will be drawn out of the carrots by osmosis, so the dry cure will become damp.)

5. When the young carrots are cured, rinse them in cool water and pat them dry with paper towels.

At least 24 hours before serving:
1. Clean and juice the large carrots, reserving the pulp to make carrot dust. (You do not need the juice for this recipe—drink it, reserve it for another purpose, or discard it.)

2. Dry the carrot pulp.
   If using a dehydrator: Place an acetate sheet in a dehydrator tray. Evenly spread the carrot pulp on the acetate sheet. Transfer the tray to a dehydrator set to 60°C (140°F) and dry the pulp, stirring occasionally, for 24 hours.

   If using the oven: Preheat the oven to 60°C (140°F) or the closest temperature available. Place an acetate sheet on a rimmed baking sheet. Evenly spread the carrot pulp on the acetate sheet. Transfer the baking sheet to the oven and turn off the heat. Let the carrot pulp dry in the closed oven for 24 hours.

3. Transfer the dried carrot pulp to a spice grinder and grind to a fine dust.

4. Weigh the carrot dust, calculate 0.1% of the total weight, add cayenne in that amount to the carrot dust, and mix well. Set aside in an airtight container at room temperature.

At least 14 hours before serving:
1. Transfer the cured young carrots to a large, heat-resistant storage container. Slice the remaining 200 g (7 ounces) peeled ginger. Put the sliced ginger in a large pot and add the orange slices, remaining 900 g (4⅔ cups) granulated sugar, remaining 35 g (2 tablespoons plus 1 teaspoon) fine sea salt, whole peppercorns, and 5 liters

(5¼ quarts) water. Bring to a boil over high heat. Pour the hot brine over the carrots and let cool to room temperature. Cover and let sit for 12 hours at room temperature.

2. Drain the carrots in a fine-mesh strainer and discard the brine.

3. Cook the carrots.
   If using sous vide: Place the carrots in a vacuum bag, seal the bag, and compress at 99%. Fill a large pot three-quarters full with water and bring the water to a boil over medium heat. Submerge the bag and cook until the carrots are tender, about 1 hour.

   If using a stovetop: Fill a medium pot with 2.5 cm (1 inch) of water and bring to a simmer. Set a steaming rack in the pot and put the carrots in the steaming rack. Steam over low heat until the carrots are overcooked inside; although the outside will look unchanged because of the skin formed by the pickling lime, the interior will be as soft as purée.

4. Dry the carrots.
   If using a dehydrator: Place an acetate sheet in a dehydrator tray. Evenly spread the carrots across the acetate sheet. Transfer the tray to a dehydrator set to 60°C (140°F) and dry the carrots, stirring occasionally, for 45 minutes.

   If using the oven: Preheat the oven to 60°C (140°F) or the closest temperature available. Place an acetate sheet on a rimmed baking sheet. Evenly spread the carrots across the acetate sheet. Transfer the tray to the oven and turn off the heat. Let the carrots dry in the closed oven, stirring occasionally, for 45 minutes.

5. Meanwhile, in a small pot, combine 100 g (scant ½ cup) water with the Pure-Cote and 0.8 g (½ teaspoon) cayenne and bring to a boil over medium heat. Add the liquid glucose and whisk to incorporate fully, about 30 seconds. Remove from the heat and set aside at room temperature.

6. Once the carrots are dried, leave the acetate sheet in the dehydrator tray or on the baking sheet. Pick up each carrot and brush the entire surface with cayenne syrup, then sprinkle with carrot dust while still wet and return the coated carrots to the acetate sheet.

7. Dry the coated carrots in the dehydrator or oven at 60°C (140°F), following the instructions above, for about 25 minutes. ❀

Spherification

Spherification is a very special, much less traditional use of charge in cooking. A truly remarkable technique, spherification was popularized by chef Ferran Adrià and quickly became known as a quintessential modernist cooking technique. A spherified dish consists of a very thin gel that encapsulates a liquid food. In the recipe in the sidebar, the liquid food is a delicious juice made from peas. We will talk about the physics of how the outside of the gel forms in the next chapter, but first let's take a closer look at the ingredients. You will notice that there are two ingredients that seem a bit unusual: sodium alginate and calcium chloride.

SIDEBAR 9:
FERRAN ADRIÀ'S SPHERICAL PEA RAVIOLI

Spherical Pea Ravioli

Ingredients

Base for the Spherical Pea Ravioli (recipe follows)
Calcium Chloride Bath (recipe follows)
Maldon sea salt
Peeled Fresh Peas (recipe follows)
Fresh Mint Oil (recipe follows)
Iberico Ham Fat Chicharrones (recipe follows)
10 pea flowers or fresh mint leaves

Directions

1. Fill a semi-spherical spoon (3 cm in diameter) with the base for the spherical pea ravioli. Drop the liquid into the calcium chloride bath. Repeat to make 10 spheres/ravioli.

2. Allow the ravioli to cook in the bath for 2 minutes. Remove the ravioli from the bath using a slotted spoon and transfer to a cold-water bath.

3. Drain the ravioli using the slotted spoon, being careful not to break them, and place each ravioli in a Chinese soup spoon. Add a small amount of Maldon salt on top of each ravioli.

4. Dress and season the peeled fresh peas with the fresh mint oil and salt to taste. Place a portion of the mint pea salad in 10 separate Chinese soup spoons. For decoration, add 2 pieces of the Iberico ham fat chicharrones and a pea flower on top.

5. Serve one spoon with ravioli and one with mint pea salad to each person.

Base for the Spherical Pea Ravioli

Ingredients

300 g frozen peas
375 g water
2.4 g sodium alginate

Directions

1. Bring the water to a boil.

2. Put the frozen peas in a blender, add the hot water, and blend into a juice.

3. Strain the juice using a fine-mesh strainer. Measure out 500 g.

4. Mix 150 g of the measured pea juice with the sodium alginate using an immersion blender until a homogeneous, clump-free mixture is achieved.

5. Add the remaining 350 g pea juice and mix well.

6. Pass the mixture through a fine-mesh strainer and store in the fridge.

Calcium Chloride Bath

Ingredients
3.2 g calcium chloride
500 g water

Directions
1. Fully dissolve the calcium chloride in the water using a hand mixer.

2. Pour the mixture into a container in which the liquid level is approximately 5 cm. Set aside.

Peeled Fresh Peas

Ingredients
80 g fresh peas
Salt

Directions
1. Bring a small pot of salted water to a boil. Add the peas and blanch.

2. Drain the peas and refresh them in salted iced water.

3. Drain the peas and peel. Store in the fridge.

Fresh Mint Oil

Ingredients
10 g fresh mint leaves
20 g sunflower oil
Salt

Directions
1. Bring a small pot of water to a boil. Add the mint leaves and blanch for 10 seconds.

2. Drain the leaves and refresh them in iced water. Remove and pat dry with paper towels.

3. Combine the mint leaves and sunflower oil in a blender and blend until smooth.

4. Press the mixture through a fine-mesh strainer.

5. Taste and season with salt, then store in the fridge.

Iberico Ham Fat Chicharrones

Ingredients

30 g Iberico ham fat

Directions

1. Remove the lean and rancid parts from the fat. Cut the fat into 0.4 cm squares.

2. In a frying pan, cook the fat squares until they are crunchy.

3. Transfer the squares to a paper towel to drain. Set aside. ⚛

Calcium chloride is a type of salt, and sodium alginate is a carbohydrate polymer—a long string of sugar molecules—that comes from seaweed. The key characteristic of the alginate polymers is that they are negatively charged, which means that they repel each other. If we are in the business of trying to make a gel, this is somewhat problematic: we need to make the polymers stick to each other, not repel each other. How could we do that? We could try adding positive charges by using table salt, which consists of positively charged sodium ions and negatively charged chloride ions. When poured into a solution of water and alginate, the ions come apart, and the positive ions will stick to the negative polymer strands and neutralize them. As a result, they will repel each other less than they did before, but they probably still won't readily stick to each other.

It turns out that there is a simple way to make the polymers stick. Just add a different type of salt, one that has two charges instead of one. A salt with two charges can neutralize the negative charge on one alginate molecule with the first charge, and

Alginate

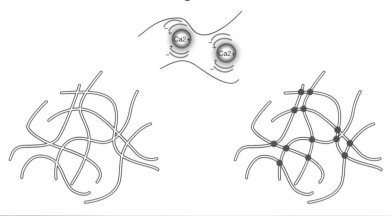

FIGURE 3 Top: Calcium ions have two positive charges, which allow them to bind to negative charges on different alginate polymers, thus binding two polymers together. Bottom: The alginate polymers (white lines) remain flexible, but the overall structure has been fixed in place by the calcium alginate bonds (orange dots), forming a crosslinked gel. Calcium is the most commonly used ion with alginate, but in theory, any ion with two positive charges could work.

then bind to another alginate molecule with the second charge. Thus, the two alginate polymers are linked together. It's as if the salt molecule were a person with two hands, each holding on to the hands of two other people, bringing them together.

In this case, we add calcium chloride, in which the calcium has two positive charges, and the alginate polymers become linked together into a network. You can't see the network and crosslinks with the naked eye, but you can see the effect of them: the liquid has changed into a solid gel. The idea is very similar to an egg that goes from liquid to a solid gel-like consistency thanks to crosslinks between proteins. We will return to the spherification recipe in the next chapter because it also does a great job illustrating another important scientific concept.

Transformations with Enzymes

So far we've discussed proteins only in the context of denaturing and changing the appearance and texture of foods, whether through heat or by other means. In these examples, the proteins have responded to changes in their environment. But proteins can also be the agents *causing* change. In fact, proteins can accomplish things with foods that would be virtually impossible in other ways.

Proteins come in many shapes and forms. Some have the ability to break other molecules apart, while others bind them together. We call these proteins enzymes. Not only are enzymes able to break and bind, but they also do it very fast—they speed up reactions that would otherwise take much longer. For example, it is enzymes that help us digest the foods we eat by breaking apart the food molecules and eventually releasing the energy we need to live. In fact, enzymes are hugely important for all living organisms, since they have the very important job of breaking apart and putting together molecules in the cell.

Up until now, we've discussed cooking as a way to deactivate proteins and make them nonfunctional. We learned about killing any potential microbes in food by pasteurization, which makes the proteins unfold and coagulate. But there are times in cooking when we want the proteins to function the way they do when an organism is alive, so we can change the food in useful ways. Recall from chapter 1 the two ways you can change food: you can change its texture or you can change its flavor. We then explained in chapter 2 that heat can do both things: it can change texture by causing the egg proteins to unravel, and it can change flavor through caramelization and the Maillard reaction.

Proteins, when functional, can also do both of these things. The flavor-creating proteins break down big molecules into small ones (just as with heat) and produce novel flavors. This is such an interesting process that we will devote chapter 7 to discussing all of the amazing flavors that can be produced this way, via fermentation.

Here we will focus primarily on the two ways that proteins can affect the *texture* of food: they can either break the food apart, or they can stitch it up together. You can imagine that your body often needs to do these things. Digesting food requires breaking proteins apart. Healing a wound requires stitching them together. Specific proteins perform these functions, which can also be used for cooking.

Proteins That Cut: Rennin, Pesto, and Pectinase

RENNIN

Many people eat cheese every day, but how many know that there is an enzyme called rennin that is critical to most cheese making? Recall that in our recipe for ricotta cheese, we added lemon juice and heat to make the milk curdle. The positive charges of the lemon juice neutralized the casein proteins so that the casein clusters fell apart. Rennin works in a similar way. It breaks one of the bonds in the casein molecules so that the negative charge is removed, which makes the casein clusters fall apart and coagulate. This is the first step of cheese production. From here you can produce a huge variety of cheeses.

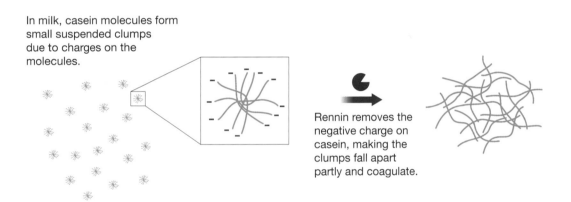

In milk, casein molecules form small suspended clumps due to charges on the molecules.

Rennin removes the negative charge on casein, making the clumps fall apart partly and coagulate.

FIGURE 4 Rennin, also known as chymosin, is an enzyme that helps young mammals digest milk. It is also a crucial component in cheese making, where its function is to cleave off a part of the casein protein. Casein is the most common protein in milk; it naturally exists in small suspended structures called micelles. The micelles are kept intact in large part by the negative charges on the casein proteins, which are distributed on the surface, as far away from each other as possible, thus stabilizing the sphere-like micelles. When the negative charge is removed by rennin, the micelles fall apart, and the casein partly coagulates into even larger clumps, which trap fat globules, ultimately making them too large to remain suspended. To the naked eye this looks like curdling, i.e., the milk separates into curds and whey. Since extraction of rennet from young calves is difficult and the yield is low, most commercial cheeses are now made with fermentation-produced chymosin, which is produced by bacteria, fungi, or yeast in which the gene for rennin has been added to their genomes. The protein is isolated after production so the modified DNA of the microbes does not remain in the final product.

PESTO

When cooking, sometimes we welcome the effect of enzymes; other times, we don't have much use for them. Have you ever made pesto and watched the brilliant green color of the chopped basil leaves turn brown and unappetizing? How can you prevent this from happening? Or, on a related note, how do you prevent cut-up pieces of apples, banana, or avocado from doing the same thing?

The browning in all of these cases is caused by specific enzymes. The enzymes are normally stored in separate compartments of the fruit or vegetable's cells. When you slice or bite it, the enzymes are released and come in contact with other molecules, as well as oxygen. The result is large brown compounds, visible to the naked eye, which we see as unappetizing browning. Normally the enzymes and the molecules are kept in separate compartments of the cell, away from oxygen, where they can't react.

SIDEBAR 10: PESTO AND PROTEIN DENATURATION

Knowing how to make proteins malfunction can be useful, as when making pesto. Pesto is a sauce made by finely chopping basil leaves, pine nuts, cheese, and garlic and mixing everything together with olive oil and salt. A common problem is that the basil, one of the key ingredients, quickly turns brown when damaged. This makes the pesto go from a rich green color to drab brown shortly after it is made. The browning is due to an enzyme, polyphenol oxidase, that reacts with phenolic compounds in the presence of oxygen. Normally the enzyme and phenolic compounds are kept apart in the cells, preventing the reaction from taking place spontaneously. It is only when the cells are ruptured while chopping up the pesto that the molecules mix and are exposed to oxygen.

Luckily, there are several ways to stop this unfortunate turn of events. Enzymes are proteins, which means that a lot of what we have already learned about proteins is also true for enzymes. Like other proteins, they have very specific three-dimensional shapes. These shapes are designed to allow them to fit certain molecules but not others, which

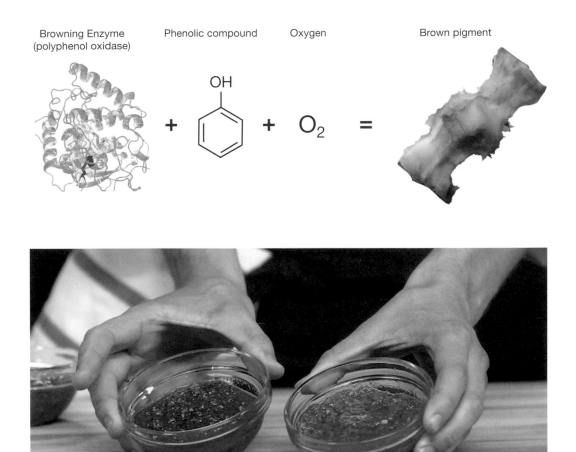

Browning Enzyme (polyphenol oxidase)   Phenolic compound   Oxygen   Brown pigment

$$+ \quad + \quad O_2 \quad =$$

gives them the ability to speed up specific chemical reactions. As with other proteins, this specific structure can easily be disrupted in the presence of heat, salt, or large changes in pH. When this happens, the enzyme no longer works.

What can you do to prevent your pesto from going brown? Try adding lemon juice or ascorbic acid—both of these decrease the pH so that the enzyme remains inactivated. You can also heat the basil leaves briefly by blanching them, which denatures the enzyme and renders it nonfunctional. ❀

PECTINASE

The examples so far have shown how enzymes are used in common, everyday foods. But given their magical powers, they're also perfect for cooks trying to push the boundaries of haute cuisine. One example comes from the creative team at Mugaritz, a restaurant in Spain's northern Basque Country, led by chef Andoni Aduriz.

The Mugaritz team has famously figured out how to use the enzyme pectinase to create entirely new dishes. Pectinase is a group of enzymes that break down the pectin. You may know pectin from jam and jelly recipes as the ingredient that gels the fruit. In nature, pectin is found between cells and in plant cell walls, where it exists as long carbohydrate chains binding the cells together. When a fruit or vegetable ripens, the pectin is broken down by a group of enzymes, collectively referred to as pectinase, and the cells detach from one another. This makes the fruit soft. The more the fruit ripens, the softer it gets, until eventually it is so soft and mushy, it's almost as if it's been cooked with heat.

Knowing this, the team at Mugaritz asked: What if you took, say, a raw apple, and injected it with pectinase? The pectinase would soften the texture, but the flavor would remain the same, right? You would get an entirely new kind of apple with a cooked texture and a tart, fresh flavor. Indeed, this is exactly what happens! You can see the recipe in the sidebar. Imagine your surprise when as a diner you are served an apple that is wrinkly, soft, and brown like a cooked apple, but when you bite into it, fully expecting the flavor of a cooked apple, it tastes tart and fresh as if raw.

The team at Mugaritz has extended this same technique to other foods. They do the same with a raw onion, cooking it with pectinase so that it becomes soft and appears cooked in consistency, while remaining completely raw in every other way. In fact, even the post-meal bad breath you'd expect from a raw onion remains the same, for better or for worse. Before serving, the onion is glazed with a delicious beef broth, creating a modern take on traditional French onion soup.

Mugaritz Apple

Ingredients
5 g pectinase
25 g water
1 red apple
Salt

Directions

1. Dissolve the pectinase in the water.

2. Poke the apple with a needle 6 or 7 times all over its surface.

3. Place the apple in a vacuum bag, add the pectinase dilution, and vacuum-pack to medium pressure.

4. Let the pectinase melt the apple for 12 hours in the fridge.

5. Open the bag and use a sharp knife to very gently cut the apple into 4 portions.

6. Serve each portion with a pinch of salt.

This recipe from Mugaritz uses pectinase to break down the pectin in apples. Since enzymes have very specific targets, only the pectin is broken up. The flavors of the raw apple are preserved. Breaking down the pectin also breaks down its cell walls and exposes the polyphenol oxidase to oxygen, causing browning. ✤

Proteins That Work as Glue: Transglutaminase

The previous enzymes all work by cutting the bonds between molecules. But enzymes can also bind molecules together, and this can be equally useful for cooking. One of the most amazing examples of this type of enzyme is transglutaminase.

If you've ever eaten imitation crabmeat, you've probably had transglutaminase. It's the ingredient that holds the paste of fish, starch, and artificial flavors together and creates a texture that resembles that of real crabmeat. In fact, transglutaminase is often referred to as "meat glue" for its amazing ability to attach different pieces of meat together. The nickname makes it sound artificial, but transglutaminase is a natural product that can be found in plants, animals, and even humans. For the purposes of cooking, it is harvested with the help of a soil bacterium, a procedure invented by a company called Ajinomoto.

Crab sticks may be a less glamorous use of transglutaminase, but a skilled chef can do amazing things with this ingredient. One of the pioneers is chef Wylie Dufresne. His restaurant named wd~50 in New York has since closed, but it was a mainstay of gastronomy for years and garnered multiple awards. Its legacy lives on in the form of the many remarkable dishes Wylie created, some of which involve creative uses of transglutaminase.

Imagine that you wanted to create noodles from shrimp. You grind shrimp into a paste, but the paste won't stick together into noodles. So, you add some binding agent such as eggs and flour. But now you have diluted the flavor. The noodles taste bland and the whole idea of calling them shrimp noodles seems useless. What to do? The answer is transglutaminase. Add it to the shrimp paste and you have perfect noodles with all the rich, nondiluted flavor of shrimp. This is what Wylie did. The final dish has the familiar shape of noodles, but a completely unprecedented flavor: shrimp.

You can even take this one step further: Want to make a novel meat by making pieces of fish and beef stick together for a marvelous novel texture? Use transglutaminase. Want to serve some oddly shaped but very expensive and not-to-be-discarded pieces of fancy meat to your dinner guests? Make them stick together into one beautiful large steak with this magical enzyme.

How does this work? To call transglutaminase a "glue" is quite a misnomer. It is no glue at all, but rather a protein like any other. It performs its magic by its ability to crosslink, or bind, two different amino acids to each other. The one type of amino acid on one piece of meat is crosslinked to another type of amino acid on the other piece of meat. There are millions of these bonds, so the meat as a whole sticks together. Moreover, the bonds are strong covalent bonds, which means they don't break even when you cook the meat. Whether or not you like the idea of eating this enzyme, we hope you'll agree that this is a pretty crafty cooking technique.

SIDEBAR 12: WYLIE DUFRESNE'S SHRIMP NOODLES

Shrimp Noodles with Smoked Yogurt and Nori Powder

Ingredients

Shrimp Noodles (recipe follows)
15 g unsalted butter
Shrimp Oil (recipe follows)
Smoked Yogurt (recipe follows)
Prawn Crackers (recipe follows)
Nori Powder (recipe follows)

Directions

1. In a small sauté pan, reheat the noodles over medium heat with the butter and a dash of shrimp oil and water, stirring and tossing as needed.

2. For each serving, paint the yogurt onto a plate and top with a generous spoonful each of warm noodles and prawn crackers. Dust the plate with nori powder.

Shrimp Noodles

Ingredients
250 g peeled, deveined shrimp
0.5 g Activa RM (transglutaminase)
3 g kosher salt
0.15 g cayenne pepper
Shrimp Oil (recipe follows)

Directions
1. In a food processor, puree the shrimp, Activa, salt, and cayenne.

2. Pass the mixture through a coarse tamis, then transfer to a pastry bag.

3. Set up a water bath to 136°F (58°C), but turn the motor off.

4. Pipe the shrimp mixture into a noodle maker or extruder and extrude it into the water bath. Cook for 2 minutes. (Heat turns on the enzyme.)

5. Using scissors, cut the noodles to the length of spaghetti and plunge them into an ice bath to cool.

6. Drain the noodles and separate them. Dress with shrimp oil and store on parchment paper lightly coated with cooking spray.

Shrimp Oil

Ingredients
200 g grapeseed oil
60 g diced onion
60 g diced carrot
60 g diced celery
10 g tomato paste
2 tarragon sprigs
60 g dry white wine or dry sake
400 g shrimp shells, chopped

Directions

1. In a large saucepan, heat about 1 tablespoon of the oil over medium heat. Add the onion, carrot, and celery and sweat until soft.

2. Add the tomato paste, tarragon, and wine and cook, stirring until tender and the alcohol is cooked down, about 10 minutes. Add the shrimp shells and the remaining oil.

3. Bring the mixture to about 200°F (93°C) and cover the pot. Remove from the heat and leave at room temperature for 3 hours, then refrigerate overnight.

4. The next day, reheat and strain the mixture through cheesecloth or a paper cone filter to harvest the shrimp oil.

Smoked Yogurt

Ingredients

225 g plain Greek yogurt
3 g sweet paprika
Kosher salt

Directions

1. Spread the yogurt in a baking pan and place in a smoker above another baking pan filled with ice. Smoke the yogurt for 3 minutes.

2. In a medium bowl, combine the yogurt, paprika, and salt. Allow the flavors to infuse for 1 hour.

Prawn Crackers

Ingredients

Neutral oil, for deep-frying
5 prawn crackers (uncooked, not the already puffed ones)
Tomato powder
Kosher salt

Directions

1. In a deep pot, heat 3 inches of oil to 375°F (191°C).

2. Crush a few prawn crackers with a mortar and pestle into irregular, smallish shapes. Deep-fry the cracker crumbs until they puff, about 1 minute. Drain on paper towels, dust with tomato powder to coat, and sprinkle lightly with salt.

Nori Powder

Ingredients

2 sheets sushi nori

Directions

1. Dry the nori sheets in a dehydrator set to 155°F (68°C) for 3 hours, or overnight in an oven with only the pilot light on.

2. Blend to a fine powder.

The enzyme transglutaminase creates strong bonds between two specific amino acids: glutamine and lysine. Depending on the type of meat, glutamine comprises up to 33% of the protein content, and lysine is also quite abundant at 8% to 10%. This means that the enzymes have a lot of material to work with. Wylie Dufresne is known for experimenting with transglutaminase to make new foods out of all sorts of proteins. Shrimp noodles is one of his signature dishes. The reaction normally takes about 24 hours to complete, but by extruding the noodles into a gentle 58°C (136°F) water bath, the process can be sped up. ❀

CHAPTER 4

Diffusion

Cooking with heat can difficult, in no small part due to the uncompromising way that heat moves through food.

—Harold McGee

We have seen thus far that foods can be cooked by a variety of cooking protocols, all leading to numerous types of phase transformations. We've learned that cookies and steak achieve their luscious texture by application of heat, ceviche cooks by absorbing acid, and pasta cooks by absorbing water. Despite their obvious differences, there is nonetheless a basic feature these foods have in common, central to their recipes. Can you guess what it is? Ceviche has a white, "cooked" outer layer and a pink, raw interior. Steak tends to have a browned surface and a very different center, ranging from rare to medium to well-done, depending on how long it is cooked. Even pasta, if you look closely, usually has a hard core surrounded by a softer consistency, at least if the pasta is cooked al dente.

For each of these foods, we can see that something seems to have happened on the outside, and this something has not (yet) happened on the inside. In each case some entity seems to have moved inward into the food, changing the texture and sometimes also the color and flavor. In the case of steak, this entity is obviously heat. Heat has cooked the outside of the meat, but the center remains mostly uncooked. In the case of ceviche, the entity is lemon juice, and the acidity of the juice gives the fish a cooked outer appearance. In the case of pasta, the entity is water. As you boil dried pasta, water slowly changes the solid pasta dough to a gel-like texture, with a core of still uncooked pasta usually remaining at the center.

But what is the mechanism by which these diverse entities—heat, lemon juice, and water—enter the insides of the food? As different as they are, they get into foods by the same process, one that is so common and crucial to cooking that it manifests itself in a wide range of recipes. Manipulating this process is the very key to a successful recipe, and the skilled chef is a master at controlling it. Without complete control of this process, pasta gets soggy, steak gets overcooked, and molten chocolate cake is no longer molten.

So, what is this mysterious process?

Diffusion to the Naked Eye

The scientific name for this all-important process is *diffusion*. When something diffuses, the naked eye can see only the effect of diffusion, not the process itself. You'll see the textural and color changes due to the heat or lemon juice, and over time you'll see these changes appearing deeper and deeper into the food.

Here's an example of what this might look like. Recall the ricotta cheese recipe that we examined in chapter 3. In this recipe, we found that adding vinegar or lemon juice to milk and heating it to 93°C (200°F) caused the milk to curdle, meaning that the proteins in the milk coagulated. What if we could zoom in and see what happens on a microscopic scale? Figure 1 shows exactly this. By placing a drop of milk and a drop of vinegar next to each other on a microscope slide and watching them mix, we have essentially made a very small amount of ricotta cheese.

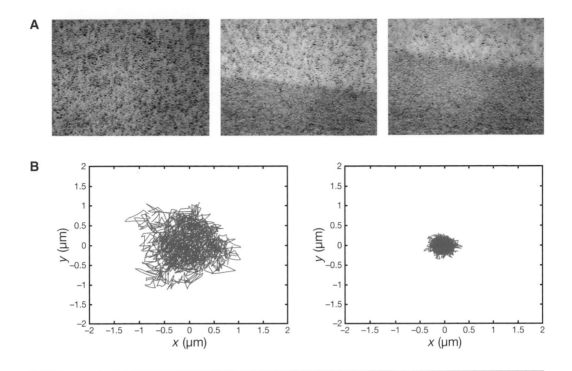

FIGURE 1 Microscopic view of the process of letting vinegar mix with milk on a microscope slide. **(A)** Left: Milk only; all the fat globules are wiggling. Middle: A front of vinegar has started moving in from the bottom, leaving behind an area that appears stationary. The top area with milk only is still wiggling. Right: The vinegar has moved almost all the way through and the entire bottom area is stationary.

(B) Left: Motion of an individual fat globule in (A) before the mixture solidifies. The line tracks the center of the fat globule, showing that it wiggles around over a wide area in a random fashion. Right: Motion of a fat globule after the mixture solidifies. The globule still jiggles randomly but is much more strongly confined in space.

If you were able to peek into the microscope before the drops mix, you would see that the milk contains lots of tiny drops that appear to move around randomly in all directions. These are fat globules. The fact that they move around is in and of itself a remarkable discovery—who'd have known that the constituents of milk wobble around like that? But in fact, all small particles in solution move. It is called Brownian motion, a phenomenon that, as we will soon see, is of great importance to this chapter.

In the next step, the milk and vinegar mix. Imagine that you keep peering into a microscope at the wobbling fat globules in the milk. As the two drops start to mix, you will see the vinegar sweeping into the picture from one side and, as it traverses inward, the movement will suddenly stop! When the vinegar reaches the milk, it becomes completely stationary. In fact, as more and more vinegar spreads through the milk, a larger and larger region under the microscope stops moving. What you just "watched" is the very idea of diffusion. As the vinegar moved in, it caused the proteins in the milk to unfold and form new crosslinks. The moving fat globules were trapped in this protein network, which prevented them from moving. We've made a gel, and you have watched the process in action.

When you think about cooking molten chocolate cake, steak, pasta, or a spherified pea juice, you should imagine the same type of diffusion process occurring. The diffusing entity enters the food, and as it works its way into the food, the impacted region grows. For molten chocolate cake, steak, and spherified pea juice, diffusion leaves behind a firmer substance than before; for pasta, the basic idea is the same but the material instead becomes softer because water itself is diffusing in.

Diffusion and Random Walks

Let's zoom in even further. What does the diffusion process look like on the molecular scale? How does the lemon juice, heat, or other diffusing entities move through the food?

If we zoom in, you will see that the essence of diffusion is the random movement of molecules. All molecules have an internal energy that makes them wiggle around slightly. When the temperature rises, molecules wiggle more; when the temperature falls, they wiggle a bit less. As each molecule does this, it is repeatedly bumped in one direction or another by surrounding molecules that are also moving. Thus, each molecule is randomly pushed in every direction. Indeed, in the preceding experiment, it was this movement that caused the fat globules to move around in the milk before the vinegar unfolded the proteins and stopped it, creating ricotta cheese.

After some time, the wiggling has the effect that the molecule travels a certain distance from its original position. Sound strange? Well, imagine that you were standing in the middle of a very large crowd of people. The people would randomly push you in all directions. Over time, you would move some distance without even trying or noticing. Something similar happens to molecules. In scientific terms, we say that the molecule has undertaken a "random walk."

History of the Random Walk: Pearson, Einstein, and Bachelier

The idea of the random walk, to our knowledge, was first introduced into the scientific literature by a great statistician named Karl Pearson. He described it in a letter to the journal *Nature* in 1905. *Nature* was then, and is today, one of the great scientific journals. At that time, it was permissible to ask questions in the articles published there, rather than propose answers as is the case today. Pearson described the drunkard's walk like this: Suppose that a person flips a coin. If the coin comes up heads, the person takes a step to the right, and if the coin comes up tails, the person takes a step to the left. Pearson called this process of randomly moving to the right or to the left the "random walk," and he asked the readers of the journal if anyone could help him understand the mathematics that underlies this process.

What Pearson didn't realize was that this question had already been answered not once, but twice, by two different people in the preceding years.

One was a young man named Albert Einstein. You might have heard of him before—he is usually remembered for discovering esoteric physics—but arguably, his most consequential paper was about random walks. (Certainly, this is his only paper that is important for cooking, at least until humans need to figure out how to cook chocolate cake inside of a black hole!) In his 1904 paper "On the Movement of Small Particles Suspended in Stationary Liquids Required by the Molecular Theory of Heat," Einstein proposed something closely related to the process we just described about making ricotta on a microscope slide, but instead of adding vinegar, Einstein just studied the milk. Recall how the fat globules of the milk were moving around randomly under the microscope before the vinegar was added. Einstein noted that the reason the fat globules move around is because they are being buffeted by the other molecules in the milk. And he asked himself what the characteristics of this process could be.

If you think about it, the process is very much like the random walk that Pearson described. Namely, imagine a fat globule surrounded by other molecules and fat globules. Sometimes the molecules on the right will push the fat globule to the left, and sometimes the molecules on the left will push the fat globule to the right. The fat globule will get pushed randomly in both directions. Einstein worked out the mathematics of this behavior in his paper, the central formula for which we will soon introduce.

The other person who answered Pearson's question is someone you have probably not heard of: Louis Bachelier, who was also a young man at the time. A student of the great mathematician Henri Poincaré, Bachelier was interested in a very different phenomenon: namely, the movement of stock prices. He wanted to make a model, a mathematical description, for how stock prices change with time. If you pick a stock and watch it, you will see that depending on various random factors, some days it goes up, and others it goes down. On the surface of it, it looks like it is undergoing a random walk, and this was indeed Bachelier's hypothesis. We now know that the movement of stock prices is a bit more complicated than this, but nonetheless, the formula he invented was a mathematical description, identical to the one Einstein had invented.

The Equation for Diffusion

In the end, Pearson, Bachelier, and Einstein invented a mathematical equation that describes diffusion. Although we have abstained from showing you equations so far in this book, this one is so important for cooking that we must share it. Perhaps we can even convince you to use it. The equation calculates the distance traveled by the diffusion front as a function of time. Without further ado, here it is:

$$L^2 = 4Dt$$

Here L is the distance that the diffusion front has moved, and t is the elapsed time since it started to move. D is called "the diffusion coefficient," and it varies for different materials. Water would have a different diffusion coefficient from heat, which in turn has a different diffusion coefficient from lemon juice. The values for different diffusion coefficients for heat are given in Figure 2. You'll see that these foods have very similar diffusion coefficients. In fact, they are all very close to the value for the diffusion of heat in water, which has $D = 0.0014$ cm^2/sec. This is because, as you might recall, most food is made mainly out of water. Our equation thus shows an amazing fact: there is a universal law for how heat travels in food that is the *same in all foods*. Imagine that.

Food	Diffusion Coefficient (x 10^{-3} cm^2/sec)
Beef	1.35
Chicken	1.36 (white meat)
Chicken	1.28 (dark meat)
Fish	1.09
Apple	1.12
Strawberry	1.27
Peas	1.82
Potato	1.23
Water	1.4

FIGURE 2 Different foods have slightly different heat diffusion coefficients, depending on their composition, but they are all pretty close to water.

What does this mean? To see, let's put our equation to the test and experiment with a recipe for molten chocolate cake. The recipe tells you to make the batter, pour it into a ramekin, and then cook the cake for about 12 minutes.

Molten Chocolate Cake

Ingredients

130 g dark chocolate chips
120 g unsalted butter (1 stick)
2 whole eggs plus 2 egg yolks
100 g sugar
60 g all-purpose flour
Pinch salt

Directions

1. Preheat the oven to 350°F (177°C). Spray 8 ramekins with nonstick baking spray.

2. In a small saucepan, melt the chocolate and butter together over low heat, stirring constantly.

3. In a medium bowl, whisk together the eggs, egg yolks, and sugar.

4. In another bowl, whisk together the flour and salt.

5. Slowly add the chocolate mixture to the egg mixture, whisking constantly.

6. Little by little, add the flour mixture to the wet ingredients and whisk well. Make sure all of the flour is completely incorporated.

7. Fill the prepared ramekins with batter so that they are a little more than half full (1.5 cm to 2 cm from the top).

8. Place the ramekins on the middle rack of the oven and bake for 12 minutes.

9. Serve warm, preferably topped with ice cream!

Molten chocolate cake, lava cake, moelleux au chocolat—regardless of what you call this dessert, the combination of soft cake with the surprise of a gooey center is one that has universal appeal. Molten chocolate cake is actually a cake that has been under-baked so that the center is still a runny batter. For this reason, this cake has been used in the Science and Cooking course since its inception to study heat diffusion. As the cake cooks, the batter around the edges reaches the temperature at which it solidifies and forms a "crumb front" that moves toward the center of the cake. If you were to take temperature measurements during the cooking process, you could calculate the heat diffusion constant of the cake batter. In the course, we use a special thermometer with multiple probes placed at different distances from the cake center. ⚛

What does our equation say about how far heat moves in 12 minutes? To find out, we need to put the values for t and D into our equation. The time elapsed is 720 seconds and the diffusion coefficient for heat in water is 0.0014 cm²/sec. Our equation then shows that the distance heat should have traveled in this time is about 2 cm, which is a pretty reasonable crumb thickness for a molten chocolate cake.

$$L = \sqrt{4Dt} = \sqrt{4 \times 0.0014 \text{ cm}^2/\text{sec} \times 720 \text{ sec}} = 2 \text{ cm}$$

At this point, you should make the chocolate cake and check out for yourself whether it works!

What is truly amazing about our equation for the diffusion of heat is that we can use exactly the same reasoning to find out how long we need to cook a medium steak and how long to fully cook the fish in ceviche. The typical recommendation for a medium steak is to cook it for about 5 minutes on each side. According to our equation, the heat diffuses 1.3 cm in this time. Double the time for both sides and you get a distance that isn't so far from the thickness of a typical steak.

For ceviche, the entity that diffuses is not heat but hydrogen ions, the molecules in lemon juice that are ultimately responsible for cooking ceviche. Hydrogen ions move at a different rate than heat. Our equation remains exactly the same, but the value of the diffusion coefficient changes. For hydrogen ions, it is $D = 0.000005$ cm²/sec. If we use our equation, we thus find that the hydrogen ions move 0.9 cm in 12 hours. If you have ever cooked ceviche so that the fish cooks all the way through, you know this is pretty accurate. Often, we cut the fish into small pieces

FIGURE 3 The image shows cubes of tuna fish that have been submerged in lime juice for 0 minutes, 30 minutes, and 90 minutes. The flesh is still red in the center. The lime juice has diffused only a short distance, resulting in a "cooked" layer of whiter fish.

before putting it in the lemon juice; this means that the ions have less distance to go and so can get there in a shorter amount of time.

Many recipes these days call for ceviche to be cooked for much shorter times. Chef Virgilio Martínez immerses the fish for less than a minute, which, according to our equation, allows the hydrogen ions to travel only 0.3 mm. They barely get into the fish at all, so the result is a very thin coating of cooked fish, almost like eating sushi.

Spherification Revisited

We described spherification in chapter 3, focusing on Ferran Adrià's recipe and the role that charged calcium ions play in sticking the alginate polymers together. But the gelation of alginate polymers is not the only reason this recipe works. At the very heart of the recipe is, you guessed it, diffusion. In fact, while spherification has lost some of its novelty in the chef community in the last few years, we still love it because it is such a marvelous example of how many different scientific concepts come together. Let's reexamine Adrià's recipe to see how.

In spherification, a delicious liquid such as olive juice, mango juice, or the pea juice from our recipe is encapsulated by a very thin gel. When you put the little gel-encased sphere of liquid into your mouth, the gel breaks, releasing a burst of wonderful flavor. The recipe starts by first dissolving very small amounts of sodium alginate into the liquid food you hope to spherify. The next step is to submerge a small spoonful of the alginate-spiked liquid food into a bath of dissolved calcium ions. As soon as the alginate blob enters the calcium chloride bath, the tiny calcium ions will start to randomly walk, or diffuse, into it. Whenever a calcium ion encounters a negative charge, it sticks. With passing time, the calcium

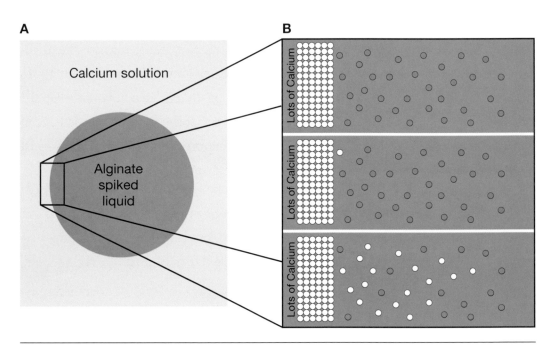

FIGURE 4 **(A)** Spherification starts when a spoonful of a delicious liquid that has been spiked with sodium alginate is submerged in a calcium chloride bath. The sphere is removed after a short period of time, at which point a thin gel has formed, encapsulating the liquid.

(B) The gel forms when the calcium ions enter the liquid by a random walk. To see how this works, we enlarge a little region of the interface of the droplet. Just as Earth looks flat to a person living on it, if we consider a small enough region, the droplet interface is perfectly flat. The top panel shows what happens right after the blob is submerged and the calcium ions (white dots) have not yet started to diffuse into the alginate spiked liquid (red dots). The middle panel shows how calcium ions start to enter the liquid. The bottom panel shows that over time, many calcium ions diffuse farther and farther into the blob. As the calcium ions move inward, they bind to the alginate polymers and form a gel.

ions diffuse farther and farther into the sphere, sticking to more and more alginate polymers. This makes the gel grow thicker and thicker.

You can see now that spherification could never happen without the process of diffusion. After all, the calcium ions cannot even begin to do their magic of linking polymers together if they do not first diffuse into the blob of food.

The secret to spherification is to ensure that the gel is just thick enough to encapsulate the liquid into a sphere, while also making sure that the gel does not become so thick that the person eating it ends up with an unappetizing gel blob in their mouth. The key to this is the key to all recipes in this chapter: understanding and manipulating diffusion.

Let's see if our wonderful equation can again help us in this matter. In this case, the thing that is diffusing is calcium, not heat, so we first need to know the diffusion coefficient of calcium ions. We know that calcium ions diffuse about 200 times more slowly than heat; the diffusion coefficient is 8×10^{-6} cm²/sec. If you wanted the gel to be very thin, say 0.3 mm, the equation tells us that the time we would have to leave the spoonful of liquid in the calcium bath is approximately 30 seconds—just about the same time recommended in the recipe for spherical pea ravioli.

$$t = \frac{L^2}{4D} = \frac{(0.03 \, \text{cm})^2}{4 \cdot (8 \times 10^{-6} \, \text{cm}^2/\text{sec})} = 30 \text{ seconds}$$

Inventive chefs have taken the basic idea of spherification and created a wide assortment of different dishes—melon "caviar," parmesan "eggs," spherified mussel juice, carbonated mojito spheres, and more. And since the taste of calcium can sometimes detract from the taste of the food, chefs sometimes use other calcium agents that work in much the same way; one popular agent is calcium gluconolactate, which has no discernable taste.

Sometimes chefs employ "reverse spherification," in which they add the calcium agent to the food and submerge it in an alginate bath, instead of the other way around. We can understand why this is sometimes helpful by thinking about the underlying science. Direct spherification, the process we described in detail above, turns out to have one important problem: Once you take the sphere out of the bath, there are still many calcium ions left on the surface of the sphere that have not yet formed bonds. These ions continue diffusing and forming crosslinks even when the sphere is out of the bath, causing the gel layer to grow thicker. Imagine throwing a dinner party, slaving over your wonderful spherified creations and tasting them when right out of the bath to check that they are perfect. When the guests arrive an hour later, the gel layer has continued to grow thicker and thicker, and you are left with a blob of gel in place of your delicious sphere. No one wants that.

The simple solution is to reverse the process—add calcium to your liquid food and immerse a spoonful of it in alginate. The crosslinking between calcium and alginate will still take place, but once you take the sphere out of the alginate bath, you no longer have a surplus of unbound alginate polymers. The crosslinking will stop, and the thin gel will remain thin for a long time—long enough for you to serve your guests, who will be very impressed with your excellent spherification skills.

WHY COOKING WITH HEAT IS DIFFICULT: A CASE STUDY OF COOKING A STEAK

We saw in chapter 2 what happens to the various constituents of food as we increase the heat from low to high. We went from adding only small amounts of heat in order to kill microbes and preserve the food, to adding moderate amounts of heat for optimal texture and color, and finally, to adding even more heat for delicious flavor in the form of browning reactions. Since some of these stages are only a few degrees apart, it is critical that you know how to manipulate the heat of each step very precisely. Even a tiny bit too much can mean the difference between a pasteurized egg, a perfect egg, and an overcooked egg.

This is just the tip of the iceberg of why it is difficult to cook with heat. The bigger problem is that you often need all of these different types of transformations to occur in the same food—and often at the same time. For example, when cooking tuna steak, you must make sure to kill the microorganisms. You also want to transform the proteins and fats in the meat while making sure not to overcook them. So far, so good. But you also want to cause browning reactions on the surface to create flavor, which occurs only at a much higher temperature. In fact, all

When the steak is "done"

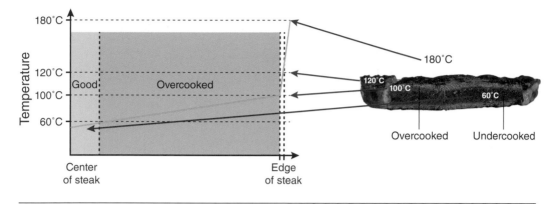

FIGURE 5 The figure shows in diagram form why it can be so hard to cook a perfect steak. The perfect temperature in the center gives the steak a reddish color. In contrast, the perfect temperature for browning reactions to occur on the surface is above 120°C (248°F). It is a basic law of physics that temperature cannot discontinuously jump from one value to another. Therefore, most of the steak has to be *between* these two temperatures, which causes these parts of the steak to be overcooked. This explains the light brown region in between the crisp exterior and the reddish interior. A major challenge of steak cooking is to decrease the size of the overcooked region.

of these important reactions occur at different temperatures. But how can we heat different parts of food to different temperatures? This is the dilemma that drives many of the protocols that we use for everyday cooking, and it is at the heart of why many cooking techniques can be so difficult to master.

A nicely cooked medium-rare steak will be less cooked in its interior, have a texture similar to well-done close to the surface, and have a crispy, brown layer right at the surface. To achieve this, we must figure out how to get the outside to be crispy without overcooking the inside. The outside of a steak requires temperatures higher than 120°C (248°F) in order for Maillard reactions to occur. The inside calls for a temperature of about 60°C (140°F), the critical temperature that corresponds to the perfect texture. The dilemma that arises from trying to balance these two requirements is illustrated in Figure 5. The steak has a very thin, brown, and flavorful crust that has been produced with the help of Maillard reactions. The center, however, is still dark pink and has a tender texture. Temperature is a continuous variable—it decreases continuously from the edge to the middle.

In order for the edge to be a perfect 120°C (248°F) *and* the center an optimal 60°C (140°F), all of the steak between the edge and the center will be between these two temperatures. So most of the steak will be overcooked. We don't want that. Chefs have designed a variety of different and creative cooking methods to deal with this problem. Let's look at a few of them and discuss how they work.

Method 1: The "normal" way. Here you heat a grill to a high temperature and deposit the steak on it. Cook the first side for 4 minutes, flip it, and cook the other side for 4 minutes. It's certainly easy, but only the very center has a perfect medium-rare texture, while the rest is overcooked. (But perhaps you like it that way!)

Method 2: Sous vide. This method works by simply cooking the steak to the exact desired temperature in a water bath of a constant temperature. If you like your steak medium-rare, this means about 60°C (140°F). The steak is sealed in a plastic bag during cooking, so no moisture is lost and the temperature is just right. This gets the perfect texture all through the middle of the steak, but it has one problem: there have not been any Maillard reactions! This can easily be fixed though by quickly searing the steak on both sides afterward. As long as the searing is brief enough, the heat will penetrate only a bit, so only the very outermost layer will be overcooked. The result: a steak with a perfect crust and almost perfect center.

Method 3: Nathan Myhrvold's method. Nathan Myhrvold invented a creative twist on the sous vide method, which completely eliminates the overcooking of the inside of the steak. After the sous vide is complete, he freezes the steak in liquid nitrogen. Liquid nitrogen is incredibly cold (–196°C/–321°F), and the rapid decrease to a very cold temperature doesn't allow ice crystals to form within the steak, thereby damaging its texture. Myhrvold then takes the frozen steak and sears it. The heat of the grill rapidly cooks the frozen steak and causes Maillard reactions on the outside surface. This same reaction happens in "normal" sous vide cooking, but the excess heat that sears the surface can penetrate the inside of the steak to overcook it. In Myhrvold's method, the inside of the steak is frozen, so much more heat would be required to bring it to a temperature where the proteins denature and appear cooked. As a result, this method cleverly gives you a sous vide steak with a delicious browned crust, but with a much thinner layer that is overcooked.

Suckling Pig with Riesling Pfalz

Ingredients

Suckling Pig Belly (recipe follows)

Black Garlic Purée (recipe follows)

Onion Purée (recipe follows)

Orange Purée (recipe follows)

Beetroot Purée (recipe follows)

Melon Cubes Soaked in Beetroot Juice (recipe follows)

Mango Terrine (recipe follows)

Black truffle slices

Purple shiso leaves

Apricot sprite hyssop flower
Suckling Pig Blanquette (recipe follows)

Directions
1. Arrange 5 cubes of suckling pig belly on each plate.

2. Add 3 dots each of black garlic, onion, orange, and beetroot purées.

3. Alternate 3 melon cubes soaked in beetroot juice and 3 cubes of mango terrine with 1 small slice of truffle on top.

4. Add 2 shiso leaves and 1 apricot sprite hyssop flower on the side.

5. Season with the suckling pig blanquette.

Suckling Pig Belly

Ingredients
2 Iberian suckling pig bellies
Brine for meat (8% salt in water)
80 g extra-virgin olive oil

Directions
1. Submerge the suckling pig bellies in the brine and refrigerate for 2 hours.

2. Vacuum-pack each belly separately with the olive oil and cook in a water bath at 63°C (145.5°F) for 24 hours.

3. Debone the bellies. Sear on a griddle with a weight on top so the skin browns evenly and turns crisp.

4. Cut into 1.5 cm squares; reserve.

Black Garlic Purée

Ingredients
50 g black garlic
70 g water
0.2 g xanthan gum
Salt

Directions

1. Blend the garlic, water, and xanthan gum, then run the resulting purée through a fine chinois. Taste and add salt as needed.

2. Reserve in a squeeze bottle.

Onion Purée

Ingredients

300 g onions, julienned
30 g olive oil
50 g unsalted butter, divided
50 g water
0.4 g xanthan gum
Salt

Directions

1. In a large skillet, toss the onion in the oil and 30 g of the butter and cook over medium heat until it softens; do not brown.

2. Using an immersion blender or food processor, blend with the remaining 20 g butter, water, and xanthan gum, then run the resulting puree through a fine chinois. Taste and add salt as needed.

3. Reserve in a squeeze bottle.

Orange Purée

Ingredients

450 g oranges, zested
400 g simple syrup
0.6 g xanthan gum

Directions

1. Fill a pot with cold water and bring to a boil. To remove bitterness, blanch the zest three times, starting with cold water each time.

2. Combine the zest and sugar syrup in a pot and cook for 15 minutes at medium heat.

3. Measure out 100 g and blend with the xanthan gum, then run the resulting purée through a fine chinois.

4. Reserve in a squeeze bottle.

Beetroot Purée

Ingredients
150 g beetroots
25 g soil distillate*
0.2 g xanthan gum
Salt

Directions
1. Bring a pot of water to a boil and cook the beetroots until soft.

2. Blend the beetroots with the soil distillate and xanthan gum, then run the resulting puree through a fine chinois. Taste and add salt as needed.

3. Reserve in a squeeze bottle.

Soil distillate is transparent like water, with taste and aroma of soil. To make it, bring water and soil to a boil, let cool, and then distill in a rotavap.

Melon Cubes Soaked in Beetroot Juice

Ingredients
1 beetroot
100 g Canary melon, cut into 0.5 cm cubes

Directions
1. Bring a pot of water to a boil and cook the beetroot until soft. Blend the beetroot.

2. Combine the beetroot purée and melon cubes and cook sous vide for 3 hours. Reserve.

Mango Terrine

Ingredients
2 ripe mangos, peeled and thinly sliced
40 g unsalted butter
3 g agar-agar

Directions
1. Preheat the oven to 320°F (160°C).

2. Place the sliced mangos in a 10 cm by 12.5 cm mold, adding some of the butter and sprinkling agar-agar in between slices. The resulting terrine should be 0.7 cm thick.

3. Bake for 30 minutes.

4. Cut into 0.5 cm cubes and reserve.

Suckling Pig Blanquette

Ingredients
0.4 g xanthan gum
200 g Clear Suckling Pig Stock (recipe follows)
20 g Cooked Suckling Pig Skin (recipe follows)
40 g Onion Oil (recipe follows)

Directions
1. Using an immersion blender, blend the xanthan gum into the suckling pig stock.

2. Add the suckling pig skin and continue blending.

3. Add the onion oil and blend until emulsified, then strain. Reserve in a bain-marie.

Clear Suckling Pig Stock

Ingredients
1 kg Iberian suckling pig bones
200 g onions, julienned
2 kg water

Directions

1. Soak the suckling pig bones in cold water for 12 hours. Drain.

2. Place the bones in a stockpot and add the onion and water.

3. Boil for 2 hours while continuously removing the foam on top, then strain to obtain a clear stock.

4. Cool and reserve.

Cooked Suckling Pig Skin

Ingredients
240 g Iberian suckling pig skin
200 g onions, chopped
150 g carrots, chopped
0.4 g xanthan gum

Directions

1. Combine the pig skin, onions, and carrots in a large pot and cover with cold water. Bring to a boil and boil for 4 hours.

2. Strain. Reserve the skin for the sauce.

Onion Oil

Ingredients
100 g onions, julienned
120 g extra-virgin olive oil

Directions

1. In a large skillet, toss the onion with the oil and cook over low heat for 30 minutes.

2. Leave to cool, then strain.

3. Reserve in a covered container. ⚛

Method 4: Heston Blumenthal's method. The final method we highlight was popularized by Heston Blumenthal. In this recipe, you flip the steak every 15 seconds on a very hot grill. If you continue flipping for several minutes, the result will be a wonderfully cooked steak. Why does this work? Every time you flip the steak, a small amount of heat browns the outside and gets into the middle. When you flip the steak, the surface cools down but some of the heat is still inside the steak and continues to cook it. By flipping the steak, the heat is continuously being deposited on both sides and is able to cook the inside without overcooking.

How can we compare these different methods? The best way of course is to buy four steaks, cook with each of the methods, and see which one you like best. This is a bit tricky for most people, requiring as it does a sous vide machine and a liquid nitrogen tank. (Recall that we showed you how to build a homemade sous vide machine in chapter 2. How to make a homemade liquid nitrogen tank remains a mystery.)

An alternative approach for comparing the recipes is to use our mathematical equation, and the extensions thereof invented by Einstein and Bachelier. To do this, we asked some students at MIT to write a software program that calculates how each of the methods for cooking the steak work. The results are illustrated in Figure 6, which shows what the interior color of the steak looks like as a function of time during three of the cooking methods. The color code corresponds to the temperature in the steak—the caption explains how this was done. The dark red is the uncooked steak, the lighter red is a rare steak (what we are trying to create), the light brown is a well-done steak, and the dark brown denotes browning reactions. You can see that browning occurs at the outside of the steak and moves inside the steak as the cooking proceeds. For the traditional method, the dark brown region goes farther inward, and even though we wanted a rare steak, part of it is still well-done. The rare steak is beautifully achieved throughout the inside by the sous vide and liquid nitrogen method. Heston's 15 seconds per side method also does a pretty good job, though there is still a big region of the steak that is well-done.

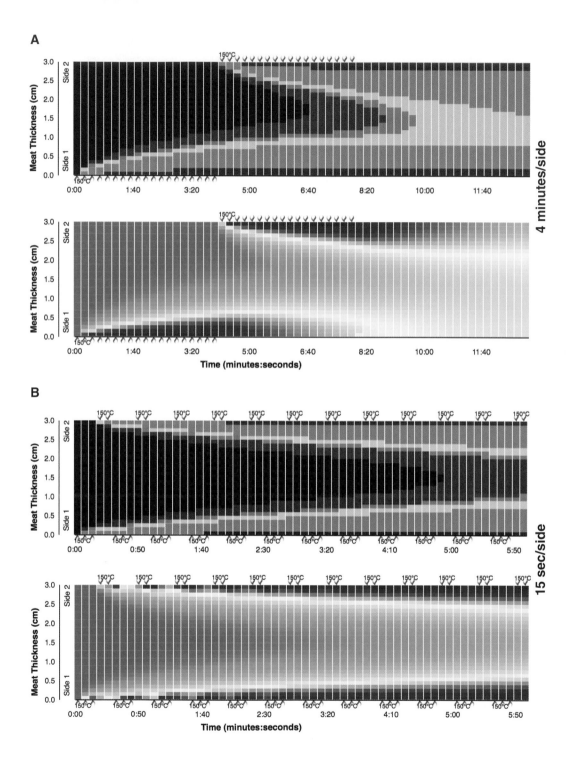

A

4 minutes/side

15 sec/side

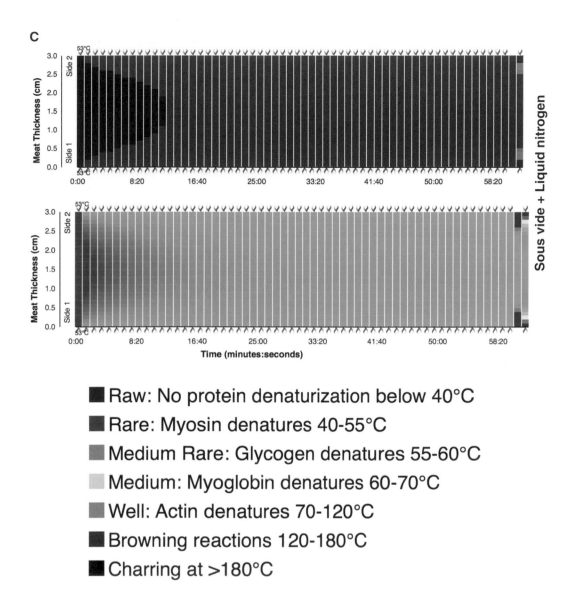

C

Sous vide + Liquid nitrogen

■ Raw: No protein denaturization below 40°C

■ Rare: Myosin denatures 40-55°C

■ Medium Rare: Glycogen denatures 55-60°C

■ Medium: Myoglobin denatures 60-70°C

■ Well: Actin denatures 70-120°C

■ Browning reactions 120-180°C

■ Charring at >180°C

FIGURE 6 Meat cooked three different ways: **(A)** four minutes on each side, **(B)** 15 seconds per side with repeated flipping, and **(C)** sous vide, followed by submersion in liquid nitrogen, followed by searing. Each cooking method is shown both in terms of the temperature profile (bottom) as well as the resulting protein state (top). The thin vertical bars within the panels correspond to what a cross-section of the steak would look like at that time point. For example, after one minute of cooking with method (A), the surface of one side of the steak is very hot, as indicated by the red, orange, and yellow colors in the temperature panel. This corresponds to a thin layer of browning, and a "cooked" appearance that is about 0.5 cm thick, as seen in the protein state panel. The final temperature and protein states, i.e., the properties of the steak when it is done, can be seen in the right-most vertical bar of each panel.

Here's one last thought about cooking steaks: Just as with spherification, where we saw that the calcium keeps diffusing after you take the spherified dish out of the calcium bath, it is the same with heat and food. You can do everything and anything you'd like to make the temperature in the center of your steak perfect. You can prepare the steak, take it off of the grill, put it on a plate, and invite your guests to sit down. Imagine that the guests then say that they'd like to go wash their hands. This takes a while, but they come back, sit down, and take a bite of the steak. Disappointingly, it turns out that by the time the guests do this, the steak has kept cooking. Air is a bad thermal conductor and the heat can't escape from the steak, so it diffuses inward! And there is no way to stop this—there is no version of "reverse spherification" for heating food. The only way to mitigate it is to remember that as the steak keeps cooking off the grill, the final temperature will change—as long as you know this, you can target this temperature appropriately.

2 minutes on each side and then let steak sit for 10 minutes.

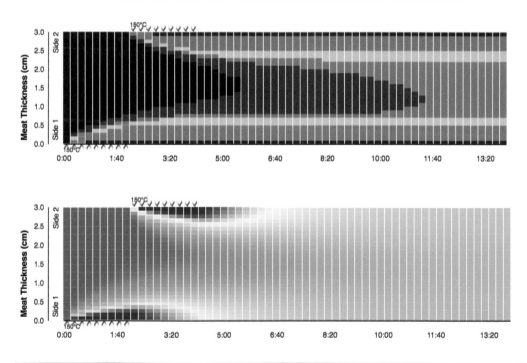

FIGURE 7 The cross-section of a steak that has been cooked for 2 minutes on each side and then left to sit for 10 minutes. The steak continues cooking and ends up perfect.

In Figure 7, we show a steak that we cooked for 2 minutes on each side, then let sit for an additional 10 minutes. The calculator indicates that this results in a perfect texture.

For dessert, let's discuss one more recipe where these principles are at play: fried ice cream! This is a dish that seems completely impossible—how can you fry something that will melt when it is heated? The secret to this seeming contradiction is, again, the careful control of diffusion. Just as with steak, molten chocolate cake, or ceviche, designing the recipe to carefully control diffusion makes it possible to achieve a dish that is perfectly cooked, both on the inside and the outside.

Here, the diffusing entity is heat, and the visual cue that heat diffusion has taken place is whether or not the ice cream melts. If heat diffuses too far, the ice cream ball will become a soggy molten mess. The name of the game is to make the outside crisp, while not letting the ice cream melt.

SIDEBAR 3: FRIED ICE CREAM

A

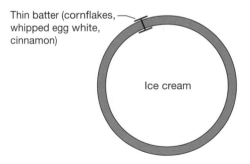

Maillard reactions

Caramelization

Frozen center

B

Thin batter (cornflakes, whipped egg white, cinnamon)

Ice cream

c

Fried Ice Cream

Ingredients

1 quart vanilla ice cream

3 large egg whites

3 cups crushed cornflakes

1 teaspoon ground cinnamon

2 quarts vegetable oil, for frying

Directions

1. Scoop the ice cream into ½-cup balls (8 in total). Place on a rimmed baking sheet and freeze until firm, about 1 hour.

2. In a shallow dish, beat the egg whites until foamy.

3. In another dish, combine the cornflakes and cinnamon.

4. Roll the ice cream balls in the egg whites, then in the cornflakes, making sure the ice cream is completely covered. Place on the baking sheet again and freeze until firm, about 3 hours.

5. In deep fryer or large, heavy saucepan, heat the oil to 375°F (191°C).

6. Using a basket or slotted spoon, fry one or two ice cream balls at a time until golden, 10 to 15 seconds. Allow the oil to come back up to temperature between batches.

7. Drain briefly on paper towels and serve immediately. ⚛

Let's deconstruct the recipe: We are supposed to scoop ice cream into balls and then coat them, first in whipped egg whites, and then in crushed cornflakes. A photo of a cross-section of the finished fried ice cream ball is shown in (A). A diagram of the ice cream ball before it is fried is shown in (B). The cornflake coating is all we have to protect the ice cream from melting. Initially this outside layer is itself much warmer than the ice cream—the eggs and cornflakes are at room temperature, while the ice cream is frozen. If we put the ball in this state into the hot oil, it would hopelessly melt.

But there is a trick. The trick is similar to Nathan Myhrvold's method for cooking a steak: Before frying, the coated ice cream ball is put back into the freezer—and left there for 3 hours. This is long enough so that when you take the ball out its temperature will be approximately that of the freezer, about – 18°C (0°F). When you place the frozen ice cream ball in the hot oil, heat from the oil starts diffusing into the cornflake coating and then into the ice cream. The temperature of the surface layer will gradually increase; after a relatively short time, it will reach the temperature at which the ice cream melts. As heat continues to diffuse farther and farther into the ice cream, more and more of the ice cream will melt.

Ideally, you would submerge your ice cream ball in the hot oil for just long enough to heat up the outer cornflake layer, turning it into a delicious golden crust, while leaving the ice cream still frozen. How long is just long enough? Well, the thickness of the cornflake layer in the recipe is approximately 0.3 cm. You can find this out by making the recipe yourself and measuring it; alternatively, you could estimate the thickness based on the dimensions of the ice cream balls and the amount of egg white and cornflakes in the recipe.

Either way, once we know the thickness of the cornflake layer, the question is how long it takes for heat to diffuse this distance. We can use our trusty equation to find out. Assuming that the rate of heat diffusion in ice cream is about the same as heat diffusion in water, we get:

$$t = \frac{L^2}{4D} = \frac{(0.3 \, \text{cm})^2}{4 \cdot (0.0014 \, \text{cm}^2/\text{sec})} = 16 \, \text{seconds}$$

So, 16 seconds is the theoretical estimate for how long we should submerge the ice cream balls in the oil. This, of course, is very close to the 10–15 seconds recommended in the recipe.

And how does this work out in practice? In our case, 15 seconds turned out to be perfect to create an ice cream ball with a golden brown crust and a still-frozen ice cream interior. The pictures in (C) show one of our ice cream balls as viewed by an infrared camera at various stages in the cooking process: right after it's out of the freezer (top left), right after it's out of the hot oil bath (top right), as it's being cut in half after cooking (lower left), and right after cutting (lower right). As you can see, in the final dish, the center (dark color) is very cold and still frozen. The surface (red/white) is very hot and deliciously browned and crispy.

A special Science and Cooking version of this recipe ensures an even thinner crust: we first dunk the ice cream balls in liquid nitrogen for about a minute. The surface layer of the ice cream ball thus gets cooled significantly—in fact, the cold diffuses into the ball in much the same way heat does. Once the extra-cold ball gets submerged in the hot oil, it will take even longer for the heat from the oil to melt the ice cream. The result is fried ice cream balls with an even thinner cooked crust, and a still-cold ice cream interior.

So, the scientific concept of diffusion is not only the secret to steak, ceviche, and molten chocolate cake. It is also the key to the intriguing hot-cold mouthfeel of this classic dessert.

Diffusion Using Brines, Marinades, and Smoking

The actions of brining and marinading are not all that different from the process of making ceviche, nor are they dissimilar from spherification. In all of these, you cook food by soaking it in a mixture of something else. The "something else" moves into the food by diffusion. Brining and marinating both involve exposing a food to some mixture of flavorful ingredients and then waiting until the flavor compounds diffuse into the food.

When you brine something, you rub it with salt or submerge it in salt water. Usually the salt concentration is high, about 20%. When brining, you wait for the salt to diffuse into the meat, and for the water to be drawn out of the cells by osmosis.

Marinades often contains salt but usually also contain many other molecules, such as sugar, acids, alcohol, and enzymes. The acids work in the same way that lemon juice cooks ceviche—it cooks and tenderizes the meat. Some fruit juices also contain enzymes that break down proteins, notably papaya and pineapple juice. Surprisingly, even ginger contains an enzyme that breaks down proteins. Marinades often contains herbs and spices; these contain larger molecules that don't diffuse very quickly, but in the long time it takes to marinate a piece of meat, they may penetrate the surface. Sometimes yogurt is added to marinades, as in tandoori chicken; the lactic acid bacteria in yogurt produce acids, which diffuse into the meat.

A typical marinade recipe will instruct that you let the food sit for 12 hours. How far would each of these compounds diffuse in 12 hours? Well, now you can calculate it.

As a cooking process, smoking also uses diffusion. Wood is made of lignin, cellulose, and hemicellulose. When you burn wood, these complex large structures break down into hundreds of small molecules of all shapes and sizes. Some are color compounds, some are taste and aroma molecules, others can act as preservatives, and yet others have an effect on texture.

Due to the high heat, the molecules vaporize and spread into the air. If you are cooking in an oven, they are bound to bump into the food. When they adhere to the surface of the food, they start diffusing into the food. The smoking and disintegrating food thus produces a molecular environment for the food that isn't so different from the calcium bath around a droplet of alginate pea juice!

During smoking, the compounds quickly react with the proteins and carbohydrates on the surface to cause Maillard reactions. This is what makes the surface of smoked foods brown. Next, they start to diffuse into the food. Small molecules diffuse faster than larger molecules. Some of the smallest molecules are carbon monoxide and nitrogen dioxide, and they end up diffusing the farthest. As they diffuse through the meat, they react with the myoglobin in muscle and change its color so you can see how far these two gases have diffused by the color change. If you are familiar with smoking food, you may have heard of the "smoke ring." Well, this is it! The smoke ring is the result of this color change due to diffusion. You can calculate the thickness of the smoke ring using our favorite equation for diffusion. Most taste and aroma molecules are larger than carbon monoxide and nitrogen dioxide, so they diffuse more slowly and don't get as far. They are colorless and don't cause a visual change in the food, so you can't tell how far they got. Given their size, it is fair to assume that they did not get farther than the distance of the smoke ring.

What smoking does to food is also temperature dependent. When you hot-smoke a food, even relatively large molecules are in gas form due to the high temperature. In cold-smoking, not as many molecules vaporize and diffuse through the food, so you get a different flavor profile.

The flavor profile is also highly influenced by the type of wood producing the smoke. Different woods have varying amounts of lignin and other constituents, resulting in an array of molecule compositions. A skilled smoker will carefully manipulate the flavor profile by controlling the temperature and humidity. The wrong temperature leads to either too much or too little decomposition of lignin, resulting in different flavor molecules.

According to several sources, the smoke ring rarely gets more than 1 cm thick. Suppose you smoke a piece of meat for 10 hours. What does this tell you about D for smoke? (Hint: Look at our favorite equation and solve for D!) You will find that D is very close to the diffusion coefficient we quoted of a calcium molecule in water. This shouldn't surprise you, because in both spherification and smoking, small molecules are diffusing in water.

Virgilio Martínez's Layers of the Amazon

We end this chapter with an amazing recipe by Virgilio Martínez, the fabulous Peruvian chef of Central Restaurante, a fixture on lists of top restaurants in the world. His "Layers of the Amazon" recipe is quite literally an artistic ode to diffusion, one that any Science and Cooking aficionado should love. Virgilio takes paiche, a white fish from the Amazon, and covers it in raw cane sugar and salt for 3 hours. Diffusion Alert #1: The sugar and salt will dissolve in the liquid and diffuse into the food, changing the flavor profile of the fish. It will also pull out the liquid from the cells, similar to what happened in the coleslaw recipe in chapter 3, giving the fish a firmer texture. Next, Virgilio creates an extract from the seed of the airampo, a beautiful cactus that grows in dry Peruvian valleys, which creates a dark red dye. He then puts the fish into the airampo extract for 1 hour (Diffusion Alert #2).

By now you know that this red dye will diffuse into the food, a distance of $\sqrt{(4Dt)}$. The picture shows the final dish. From the streaks of red color and your knowledge of diffusion, you can calculate the diffusion coefficient of the dye! A science experiment from a beautiful recipe—what could be better?

Layers of the Amazon

Ingredients

Copoazu Cream (recipe follows)
Yuca and Achiote Crisp (recipe follows)
Airampo-Dyed Paiche (recipe follows)
20 bahuaja nuts (Brazil nuts)
20 alfalfa flowers (*Medicago sativa*)

Directions

1. Put some copoazu cream in the middle of the plate.

2. Cover with the yuca and achiote crisp.

3. Top with the airampo-dyed paiche.

4. With a thick Microplane, shave the bahuaja nuts. Place the shavings in 5 different blank spots.

5. Garnish with the alfalfa flowers.

Copoazu Cream

Ingredients

200 g copoazu pulp (*Theobroma grandiflorum*, related to chocolate)
50 g raw cane sugar
170 g heavy cream
2 gelatin sheets, bloomed (soaked in cold water for 5 to 10 minutes and drained)

Directions

1. In a small saucepan, cook the copoazu pulp with the sugar over medium heat for 10 minutes, or until it is dissolved.

2. Let it chill, then blend with the heavy cream until the fiber of the copoazu is broken down.

3. Pass the mixture through a strainer and heat the mixture again until it reaches 158°F (70°C).

4. Add the dissolved gelatin and incorporate well.

5. Transfer the mixture to a 1 L siphon with 2 nitrous oxide charges and chill.

Yuca and Achiote Crisp

Ingredients

700 g vegetable stock
40 g achiote seeds
60 g tapioca starch (extract of yuca)
50 g huito pulp (fruit of the genip tree, *Genipa americana*)

Directions

1. Heat the vegetable stock in a small saucepan over medium-low heat. Add the achiote seeds, and let them infuse and dye the stock red.

2. Let the broth cool, then add the tapioca starch and huito pulp.

3. Heat over medium heat, stirring frequently, until the tapioca is completely dissolved and the mixture is dense and thick, about 10 minutes.

4. Spread 200 grams of the mixture in a thin (1 mm), uniform layer.

5. Dehydrate for 12 hours at 140°F (60°C).

6. Deep-fry at 392°F (200°C) to puff up the crisp.

Airampo-Dyed Paiche

Ingredients
1 kg salt

1 kg raw cane sugar

500 g paiche fillet (*Arapaima gigas*, an Amazonian fish)

500 g airampo extract (*Opuntia soehrensis*, a cactus with edible pink fruits)

Directions
1. Combine the salt and cane sugar and cure the paiche in the mixture for 3 hours.

2. Remove the paiche from the dry cure and rinse with cold water.

3. Submerge the paiche in the airampo extract and leave for 1 hour, or until the airampo dye penetrates the outside of the paiche. ✿

CHAPTER 5

Texture, Viscosity, and Elasticity

When we eat, we put food in our mouth, and we then apply force to that food. We use our teeth to break things up into small enough pieces to swallow. And we use the tongue to move things into place for chewing and swallowing, but also to perceive and interpret the physical behavior of the food, which could be all sorts of different textures. If the food is liquid, then it could be thin or thick or slimy. It it's solid, it might be hard or soft or crumbly or malleable or stretchy. Foods are generally combinations of all of these different qualities. So the mouth is a very sensitive analytical instrument for food structure.

—Harold McGee

What Is Texture?

Texture is incredibly important when it comes to food—often just as important as flavor. As we've learned thus far in this book, the two are so closely intertwined that it can be difficult to separate one from the other. They seamlessly blend together and make for a perfect bite. If one is off, we dismiss the dish in its entirety. Most of us can think of times when a food had excellent flavor but was still disappointing because the texture was off in some way. Just imagine a soggy French fry—you get the point.

What would a food taste like if its flavors were gone and the only thing left was texture? It's hard to imagine, but here is a way you can actually experience it: go to a health food store and get the powder from the *Gymnema sylvestre* plant. Coat your tongue with the powder. It's not very tasty, so don't take too much. Next eat some sugar. What is it like? For most people the sugar now tastes like sand. The sweetness is gone because a compound in the herb has blocked your sweet receptors. The only thing left is texture. Would you eat sugar if it tasted like that? Probably not. You might as well go to the beach and eat some sand, as it would be less caloric! Nevertheless, from a scientific perspective, this experiment illustrates what we plan to do in this chapter: look at texture as isolated from other aspects of foods.

Chefs are highly skilled at balancing flavor and texture and then playing them off each other in interesting ways. If eating sand made you less interested in knowing about texture, take a look at this amazing creation by Jordi Roca, whose work we encountered earlier in the book. The dish has a variety of textures, all with the same basic apple flavor. Take a moment to admire it. Imagine how boring it would be if all the flavors in your life came with exactly the same texture. Almost all outstanding recipes in haute cuisine have this kind of diversity in texture, making for more interesting food. If you examine Jordi's recipe, you will notice that each texture is made in a different way. Our goal in this chapter is to explain how these work. At the end of the chapter you should be able to return to this recipe and deconstruct it on your own.

Jordi Roca's caramel apple dessert features a variety of textures all with the flavor of apple: the brittle caramel apple itself, fresh apple, sautéed apple, an airy foam, a gel, a compote, all brought together by the cold sensation of ice cream.

Caramel Apple

Ingredients

Apple Compote (recipe follows)

Apple Sautéed with Calvados (recipe follows)

Apple Jelly (recipe follows)

Royal Gala apple, julienned

Caramel Apple (recipe follows)

Apple Foam (recipe follows)

Calvados Ice Cream (recipe follows)

Directions

1. Put a base of apple compote on a plate.

2. Arrange some sautéed apple balls, apple jelly cubes, and fresh apple julienne on the sides.

3. Fill the caramel apple with apple foam. Set it on top of the compote, along with a quenelle of Calvados ice cream.

Apple Compote

Ingredients

750 g Royal Gala apples, quartered
75 g sugar
150 g unsalted butter, at room temperature

Directions

1. Preheat the oven to 392°F (200°C).

2. Combine the apples and sugar in a small baking dish and bake until very soft.

3. Transfer the apples and sugar to a blender, add the butter, and puree.

Apple Sautéed with Calvados

Ingredients

400 g Golden Delicious apples, peeled
40 g sugar
1 vanilla pod
20 g Calvados
30 g unsalted butter

Directions

1. Using a 1 cm melon baller, scoop out balls of apple pulp.

2. Melt the sugar in a saucepan over low heat until it turns to caramel.

3. Add the vanilla pod and the apple balls and cook over medium heat until soft.

4. Pour in the Calvados and flambé.

5. Whisk in the butter.

Apple Jelly

___

Ingredients
300 g natural apple juice
3 gelatin sheets, bloomed (soaked in cold water for 5 to 10 minutes and drained)

Directions
1. Heat 100 g of the apple juice in a small baking dish, add the gelatin and set aside until it melts.

2. Add the remaining 200 g juice, mix well, and transfer the mixture to a container to solidify.

Caramel Apple

___

Ingredients
250 g fondant
125 g isomalt
125 g glucose
10 drops 50% citric acid solution
4 drops red food coloring

Directions
1. Combine the fondant, glucose, and isomalt in a pot.

2. Heat over medium heat to 320°F (160°C), then lower the heat to reduce the temperature to 302°F (150°C) and add the drops of citric acid and red coloring.

3. Pour the sugar mixture quickly over a silicone mat and stretch and fold over approximately 20 times until it turns satiny.

4. While hot, form the mixture into approximately 2 cm balls.

5. Heat the tip of a blowing pump, insert it into a still-warm ball, and blow to form a ball of the desired size.

6. Shape the ball like an apple and reserve in an airtight container with silica gel, or in a cupboard for sweets.

Apple Foam

Ingredients
750 g Golden Delicious apples, quartered
75 g sugar
60 g unsalted butter, at room temperature
250 g egg whites

Directions
1. Preheat the oven to 284°F (140°C).

2. Combine the apples and sugar in a small baking dish and bake for 35 minutes.

3. Transfer the apples and sugar to a blender and blend.

4. Pass the mixture through a sieve and add the butter while mixing vigorously to a smooth purée. Set aside to cool.

5. Beat the egg whites and fold in the apple purée.

6. Transfer to a 1 L siphon with 2 nitrous oxide charges and reserve.

Calvados Ice Cream

Ingredients
541 g whole milk
171 g heavy cream
47 g skim milk powder
176 g sucrose
9 g cream stabilizer
5 g buttermilk protein powder or skim milk powder
51 g Calvados

Directions

1. Mix the milk, cream, and milk powder in a saucepan and heat to 104°F (40°C). Add the sucrose, stabilizer, and proteins.

2. Heat to 185°F (85°C), then cool to 39°F (4°C).

3. Add the Calvados and let sit for about 8 hours.

4. Run the mixture through an ice cream machine and keep at 0°F(−18°C). ❀

Generally, food appears in the form of a liquid or solid. Food is rarely served as a gas. So when we want to describe the texture of certain foods, we use words that describe these two types of materials. We can do this in numerous ways, but there are two descriptors that capture the general idea.

For liquid food, we want to know how *thick* it is: Is it thick like a milkshake, viscous like honey, or thin like water? We refer to this property as viscosity.

For solid food, we want to know how *firm* it is: Is it hard like rock candy, squishy like a steak, or soft like a chocolate mousse? We refer to solid foods as having an elastic modulus (aka Young's modulus); the higher the elastic modulus, the stiffer the food.

Whether you're a chef or a home cook, much of what you do when cooking is manipulating these two properties. As it turns out, the microscopic principles underlying both viscosity and elastic modulus are very similar and can be grouped into two governing principles. A food's texture is always determined by one of these, and sometimes by a combination of both.

The first principle is one we have previously touched on, namely the idea that polymers of food—proteins, carbohydrates, or other very long molecules—can form a network that changes texture. We saw this in the phase transitions that occurred when protein polymers in eggs denatured and coagulated from heat, transforming the egg from liquid to solid. We also encountered it in spherification, where the alginate polymers formed a solid gel that encapsulated the blob of liquid. There are many polymers that can increase the viscosity of liquid foods, making them change from watery to having the thickness of a milkshake. If you add enough polymers and the network becomes sufficiently strong, the liquid can become so thick that it becomes a solid. You can take this even further, and make an already-solid food even stiffer by increasing the strength of an existing polymer network. The main point is, you can modulate both elasticity and viscosity with polymers. We will return to this idea soon.

The second of the two principles governing food texture is an idea that we will refer to as the "packing" principle. Just like polymer networks, packing also determines the properties of both liquid and solid foods, changing both viscosity and elasticity.

Packing

What exactly is packing? You are well aware of packing in your daily life—you pack your clothes to go on a trip, you pack your fridge with groceries, and so on. You might be surprised to find out that this same idea is critically important in cooking. Controlling how well ingredients pack together is essential to controlling texture. By changing this single "knob," we can dramatically change the texture of a dish.

To explain this, let's begin with a riddle: Suppose you go shopping at the grocery store and when you arrive home, you realize to your horror that there isn't enough room in the pantry to store all of your food. What are you to do? An organized person might try to reason their way through this—they could start by placing the largest items on a shelf and then filling in the gaps with the smaller items, hoping that everything will fit. A less organized person might just play it by chance and stick things on the shelf in random order, hoping that there is enough space for everything to fit. Which method do you think will work better?

This is a packing problem, for you are trying to figure out how to arrange non-overlapping objects in a three-dimensional space to get as much in as possible. Mathematicians love packing problems—what can be more fun than figuring out what will fit and what won't?

As we all know from real-life packing situations, the most important factor is how much stuff we need to jam together. The critical comparison is the ratio of the volume of our objects to the volume of the container; the term for this is *volume fraction*. The volume fraction determines whether you can randomly throw items in or if you need to agonize over exact placement. Of course, if the total volume of the objects is larger than the total volume of the container, then there is no way they can all be packed in no matter how skilled you are. The more things we try to pack into one place, the more crowded it gets and the fewer ways that objects can be arranged without overstepping the boundaries. In contrast, the emptier our pantry, the more freedom we have: with a spacious pantry and very few grocery items, it is extremely easy to reach in and move or take out particular items without needing to disturb anything else. However, if the space is barely large enough to contain all the items, we may instead need to arrange them in a particular way based on their size and shape so that the door can actually close.

Now, what does this have to do with food and cooking? We know that all foods are made up of many molecules and chunks of their ingredients. These molecules and chunks are distributed throughout the food, and the free space between them is filled mostly with water or air. Most importantly, the texture is controlled by how tightly the ingredients are packed together. The higher the volume fraction, the firmer the food.

To illustrate this, consider the recipe for a basic marinara sauce in the sidebar. Let's go through the recipe in detail and figure out where packing comes in. The first step in the recipe is to sauté onion and garlic in oil to extract its flavor molecules, which are not water soluble. Then we crush whole tomatoes into tiny pieces, extracting the juices, and add even more water, diluting the tomato pieces further. All of this is added to the onion and garlic. With all the extra water, the tomato pieces are not so densely packed, and the sauce is thin and flows easily. To thicken the sauce, we simmer. The idea is to evaporate some of the water while at the same time concentrating the flavor—the flavorful tomato chunks don't evaporate, since they are too heavy. As the water evaporates, the tomato chunks become more and more crowded and the sauce becomes thicker and thicker. Finally, the tomato chunks are packed pretty tightly together; at that point the sauce has thickened enough and is ready to serve. The process of adding water and boiling it off allowed us to make a thicker sauce and at the same time concentrate the flavor by having more tomato chunks per bite.

SIDEBAR 2: MARINARA SAUCE

Marinara Sauce

Ingredients
1 (28-ounce) can whole San Marzano tomatoes
1 cup water
¼ cup olive oil
1 medium onion, thinly sliced
8 garlic cloves, slivered
1 small dried red chile, cut in half, or crushed red pepper to taste
1 teaspoon kosher salt
1 large fresh basil sprig
¼ teaspoon dried oregano

Directions
1. Pour the tomatoes with their juice into a large bowl and crush by hand or with a spoon. Add the water to the can to get any remaining tomato juices and reserve.

2. Heat the oil in a large skillet over medium heat and add the onion. When the onion is just softened, add the garlic. As soon as the onion and garlic are aromatic but not browned, add the tomatoes, then the reserved tomato water. Bring to a slow simmer, then stir in the chile or crushed red pepper and the salt.

3. Add the basil and allow it to wilt before stirring it in. Stir in the oregano.

4. Simmer the sauce until thickened, about 15 minutes. Remove and discard the basil and chile.

5. Serve warm with pasta or use with pizza. Refrigerated leftovers will keep, covered, for up to 4 days. ❀

If instead of serving the sauce, we keep simmering it beyond this point, even more water evaporates and eventually the sauce becomes tomato paste. This is an entirely different substance: a chunk of tomato paste placed on a dish will maintain its shape and will not flow. Lo and behold, by getting rid of enough water, our tomato sauce has become a solid. This solidification is due to the very tight packing of the tomato particles. When enough objects have been packed into a space, they can reach a point where the packed objects are stuck against each other and can no longer move. Scientists say that the material has become *jammed*. Once it is jammed, the substance becomes a solid. Simply by increasing the packing of tomato chunks in water, we've gone all the way from a very thin liquid to a more viscous liquid, and finally all the way to a solid—simply by increasing the number of tomato chunks per volume of water. This is the very essence of how packing controls texture in foods.

Indeed, the liquid turned into a solid. But *when* exactly did this happen? Is there a magic number of tomato chunks at which point the sauce becomes a solid? Indeed, there is. To get to it, we must spend a little more time thinking about tomato chunks.

Volume Fraction

In both the thin sauce and the tomato paste, the amount of tomato does not change, but the amount of water—and the total volume of sauce—does. A thin marinara sauce has fewer tomato chunks per volume of water than a thick sauce. We can say that its *volume fraction* of tomato chunks is lower.

The volume fraction is the ratio of the stuff (usually the solid ingredients) you are mixing together in your recipe compared to the total volume. The total volume is everything—that is, the sum of the volumes of all the stuff you are mashing together *plus* the volume of pure liquid (usually water). Pure liquid provides room for the "stuff" to move around. When a recipe modifies the volume fraction, you are manipulating the texture of the food. Besides the marinara recipe, another place we have seen this is in the candy recipes in chapter 2. There, as the temperature went up, the water evaporated from the sugar and the texture of the candy changed.

This is a very simple idea, but don't let that fool you: it is also quite profound and will help you understand a lot about how recipes work. This is the equation that summarizes the idea:

$$\text{Volume Fraction} = \frac{\text{Volume}_{\text{stuff}}}{\text{Volume}_{\text{stuff}} + \text{Volume}_{\text{water}}}$$

Writing it this way makes it clear that we have two ways to manipulate the volume fraction: either change the volume of stuff ($\text{Volume}_{\text{stuff}}$) by adding more or less of it, or change the volume of water ($\text{Volume}_{\text{water}}$) that the stuff moves around in. This seems really simple, but it is not as simple as it appears. In both the marinara recipe and the candy recipes, we changed the volume fraction by simmering the mixture and boiling off the water. This increased the volume fraction because we decreased the amount of water. Conversely, we could simply increase the volume of tomato chunks or sugar added.

Now we can return to the question of when the marinara sauce became a solid. The answer to this is pretty universal and doesn't depend on what you are cooking. The same answer applies to tomato particles in a marinara, flour in a cup, or soy particles in tofu: the rule of thumb is that jamming happens when the volume fraction of the stuff is between 65% and 70%. Up until this point, the tomato sauce is viscous, it flows, and in order for it to become a solid, you need to boil off enough water so the tomato chunks occupy more than 65% of the total volume. This has important implications for cooking. It also turns out to be useful in other critical aspects of your life, such as winning the popular party game of guessing how many jelly beans are in a jar. When the jar is full and the beans are no longer jostling around, you can be certain that the volume fraction is around 65%–70%. So this means that about 30% of the volume of the jar of jelly beans is filled with air! See the sidebar on M&M's to explore this, and the concept of packing, in more depth.

SIDEBAR 3: M&M'S IN A JAR

How many jelly beans are in the jar? Knowing about volume fraction and jamming, we now know a way to estimate it. In this case, since the jelly beans do not pack perfectly because of their shape, roughly 70% of the container's volume is filled by candy. The rest is air in the space between jelly beans.

You can explore how shape affects packing by filling jars with other kinds of candy. Which shape packs more densely? Spheres, like gumballs, pack less efficiently than any

Jelly beans

Gum balls
(spheres)

M&M's

other shape—worse than, say, M&M candies and jellybeans. The reason is that other shapes can rotate and squeeze their smallest side into the available spaces. Notice the different orientations of the M&M's in the jar on the right. But no matter how you rotate a sphere, like a gumball, it is still the same size. You can confirm this for yourself by doing the following experiment: Fill two jars with two different kinds of candy and put each jar on a scale. Tare the scales so they adjust to zero. Then pour in water until you've covered all the candies—in other words until all of the air spaces in between have been exchanged for water. The scale will tell you how much water you added, and by knowing the total volume of the jar you can find the volume fraction of the candy. See if you can confirm that spheres require more water! Just don't count on eating the candies afterward—since candy is mostly sugar and sugar dissolves well in water, your colorful candies will likely transform into an unappetizing brown goop. ⚛

Before we move on from the idea of packing, let's stop to consider one more example. As it turns out, you don't have to go through the lengthy process of cooking marinara sauce to encounter this principle in the kitchen. In fact, all you have to do is measure out a cup of flour. Flour, as you know, is made of tiny flour particles. When you scoop into the bag to fill your cup, the particles will pack into it, but how much flour you get is entirely dependent on how closely the particles pack. Take a look at the sidebar for how flour amounts can vary as people around the world did this exercise.

SIDEBAR 4: PACKING FLOUR

You would think that "1 cup of flour" is a highly specific measurement, being widely used in recipes, but this is not quite the case. It can mean widely different things depending on how densely the flour particles are packed. In our online class, we asked nearly three thousand people from around the world to measure out a cup of flour and then weigh it. The diagram shows the range of weights. Most people got 130 g–160 g of flour in a cup, but some get less than 90 g, and others more than 250 g!

Weight of 1 cup of flour as measured by course participants (*n*=2717)

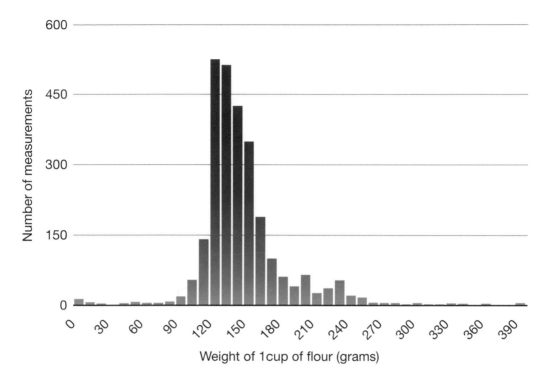

Cookbooks, and baking books especially, often give the advice to weigh the ingredients in the recipes instead of measuring by volume. For example, the King Arthur Flour Company states that for their recipes, 1 cup should be interpreted as 120 grams. If you've ever ignored this advice, this diagram may make you think twice, since a cup of flour can mean different things depending on who is doing the measuring. It all depends on the packing. You may have noticed that by tapping the measuring cup on the counter as you fill it, you can fit more and more flour into the cup, even if it initially looked full. The small motions generated by the tapping help the granules rearrange into a more compact arrangement. If you're measuring by weight, you always get the same number of flour particles regardless of how densely they are packed. Lucky for us, though, many recipes work out anyway—they are robust enough that the exact amounts don't matter. ✺

You now have a sense for the two main principles that determine texture in liquid and solid foods: manipulating the packing and manipulating the network formed by various polymers. More details will emerge below and, as is often the case, the beauty is in the details. Packing can happen in many ways depending on the ingredients and the circumstances under which it occurs. Similarly, networks can have different properties depending on the type of polymer. Often, a food's texture is due to a combination of both principles. Nevertheless, you will see that these two ideas are at the heart of most of the transformations we see in cooking. They are the two main control knobs we have at our disposal for manipulating texture in foods. Figure 1 summarizes what these ideas look like on a microscopic scale.

Now, let's look at the details. We'll start with viscous materials first, then move on to solids.

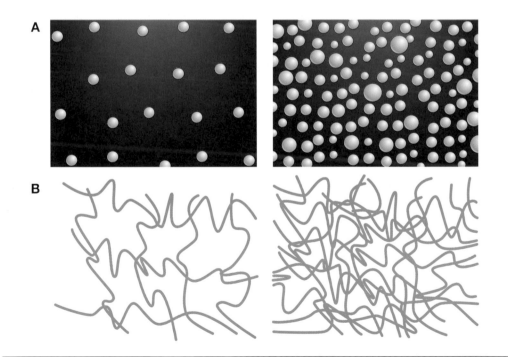

FIGURE 1 The two main contributors to texture are packing and polymers. The more dense the packing or polymer network is, the thicker or firmer is the resulting material.

    (A) Packing with sparse particles (left) and dense particles (right).

    (B) Packing with a sparse polymer network (left) and a dense polymer network (right).

Viscosity

A liquid is a substance that can flow and fill a container. The concept of viscosity is what quantifies how *quickly* the liquid flows or, put another way, its ability to resist flow. A fluid with high viscosity flows slowly, while a low-viscosity fluid flows quickly.

As you may suspect, the very essence of this concept can be understood through packing—in this case, it's the packing of molecules, not of the tomato chunks from before. If you try to wade through a ball pit, the plastic balls need to move away and over each other in order to let you pass. In a pool of water, the water molecules must also move aside for the water to flow around you. If the balls or water molecules can't easily move, then flow is more difficult and viscosity increases. There are two ways that this can occur. The first is when the molecules that make up the fluid become larger or more complex, which makes them flow over one another more slowly. Such is the case for oil, which flows more slowly than water, and hence is more viscous. The molecules that make up the oil are much larger than water molecules, and they flow over one another much more slowly, increasing the viscosity as compared to water. The second case occurs when particles or molecules are added to the fluid, which is almost always water. When the particles or molecules are closely packed, it is more difficult for them to move over one another, and the viscosity increases. The more easily the pieces can move around each other, the more easily the material flows, and the lower its viscosity. This implies that the higher the volume fraction of stuff, the higher the viscosity. Milk feels slightly more viscous than water because it has tiny drops of fat suspended within it. The fat drops can still easily move around, which is why milk pours relatively easily. Our tongues are quite sensitive; whole milk is only 3%–4% fat, yet it is enough for us to feel a difference between it and water. Heavy cream, which contains 36% fat or more, is even more viscous.

Five Ways Chefs Increase Viscosity

Now that we know what viscosity is, how do we manipulate it when we are cooking? It seems simple: we must either change the volume fraction of the stuff in our food, or create and manipulate some network of polymers. Lowering the viscosity of a recipe is easy—just add water! (Note that no liquid we eat has a viscosity lower than water.) The problem with this method, however, is that it dilutes flavor. If you take your tasty but too-thick sauce and then just pour water in it, you will mute the flavor. For this reason, cooking methods are concerned with ways of changing the viscosity of a dish while maintaining or even increasing flavor. There are myriad ways of doing this. All of the methods can be grouped into five main categories: (1) reduction, (2) adding "stuff," (3) making emulsions, (4) making polymer networks from food components, and (5) using modernist thickeners. Let's delve into the science behind each method.

REDUCTION

Reduction intensifies flavor. The method involves increasing the concentration of particles in a fluid by boiling off some of the liquid until the desired thickness is reached. This is thickening by increasing the packing density of the molecules or particles in the sauce. The marinara sauce was an example of this. Another is reducing wine for a red wine sauce. One disadvantage of this method is that it can take a very long time to boil off enough liquid for the food to become thicker. In the case of a red wine sauce, for instance, because there are not very many particles in wine to start with, it is not uncommon that 90% of the liquid must be boiled off for substantial thickening to occur; this, of course, takes time. If a liquid contains even fewer particles than wine, then the method of reduction is not very effective at all at increasing viscosity.

ADDING STUFF

The second way to thicken a liquid food is by adding stuff. We can add all types of things, like tomato chunks, or sugar as we did in the praline recipe, or really anything. As long as it increases the volume fraction, it increases the viscosity.

Sometimes cooking itself can change the volume of the stuff! Let's illustrate this with a classic recipe for risotto.

Risotto

Ingredients

4½ cups chicken broth

3 tablespoons olive oil

4 tablespoons unsalted butter, divided

¾ cup minced onion

1 leek (white part only), trimmed and chopped

4 scallions, chopped, whites and greens reserved separately

2 cups Arborio rice

⅓ cup dry white wine or vermouth

⅓ cup grated Parmigiano-Reggiano

1 tablespoon chopped fresh parsley

Salt and ground black pepper, to taste

Directions

1. Heat the broth to a simmer in a medium saucepan. Turn the heat to low, just enough to keep the broth hot.

2. Heat the oil and 2 tablespoons of the butter over medium heat in a large, heavy-bottomed saucepan. Once the butter has melted, add the onion and cook until softened, about 4 minutes. Stir in the leek and scallion whites and cook until the onion is golden, about 6 more minutes.

3. Add the rice and stir until toasted, 1 to 2 minutes.

4. Add the wine and cook until the wine is fully absorbed.

5. Add a ladleful of the hot chicken broth, enough so that the rice is just barely covered, and continue cooking and stirring until the liquid is absorbed.

6. Repeat step 5 until the rice is fully cooked, 16 to 20 minutes. The rice should be tender but still firm.

7. Remove the pan from the heat and stir in the remaining 2 tablespoons butter, half of the Parmigiano, the scallion greens, and the parsley. Season to taste with salt and pepper.

8. Serve immediately in bowls, passing the remaining Parmigiano for garnish.

The iterative process of making risotto, by adding a little liquid at a time, allows you to control the texture of the rice and sauce. As liquid is added, the rice releases starch, which thickens the sauce (by increasing the volume fraction). ⚛

Chef Lidia Bastianich made risotto when she visited our class. She used a classic technique for making risotto, similar to that in the sidebar recipe. Classic risotto starts with toasting the rice, then slowly adding hot broth, just enough to cover the grains, and stirring the mixture constantly. After all of the liquid has been absorbed and boiled off, more liquid is added ladle by ladle, and this cycle is repeated until the rice is cooked through. It's a lengthy process, perfect for relaxing or socializing over a glass of wine—or, for that matter, discussing science, which is what we did in class while the delicious aroma filled the lecture hall and made us all very hungry.

Rice has an amazing ability to absorb water. As each cup of broth is added, the rice slowly swells. This increases the volume of the rice overall. In essence, you're thickening the risotto by letting the stuff that is already there expand. What's more, the rice releases starch into the liquid. Starch comes in very small granules, and as these granules are released by the rice, they swell and make the liquid even thicker, thereby adding more stuff. There are a few reasons that the recipe calls for adding only a little bit of liquid each time. One is that by slowly adding the broth, the starches stay well distributed and thicken effectively rather than clumping up. Another reason is that batches of rice are not all the same. The starch content, age, and hydration levels depend on the particular batch and its storage conditions. By slowly adding the broth, you have great control over the final texture. It's easier to add more broth than to thicken after the rice has reached al dente firmness.

In this recipe, there are three reasons why the volume fraction of "stuff" increases while cooking. First, the volume of the rice increases since the heat causes it to swell. Second, the water evaporates off. Third, the starch granules from the rice swell *a lot*—remarkably, the radius of a starch granule can expand by a factor of three. This means that the volume goes up by a factor of 30. The effect is even larger than

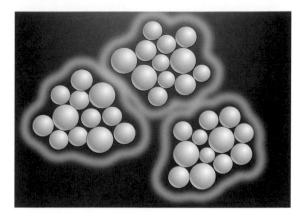

FIGURE 2 Starch is a powerful thickener for several reasons: Not only do starch granules swell when hydrated and heated, thus taking up a larger volume, but they also stick together in tenuous structures whose effective volume fractions include both the granules and the enclosed solution. This increases the volume fraction further. Finally, the starch granules also leak polymers, which increase the viscosity further by creating a polymer network (not shown here).

this: starch granules have a tendency to clump together into tenuous structures. The volume of these structures is the volume of the structures themselves *plus* all the water that they enclose. This further increases the effective volume fraction of the particles. The combined iteration of all of these factors creates a creamy, wonderful risotto.

The magical ability of starch to thicken liquids is a useful trick and widely used in the kitchen. There are many versions of starch-based thickeners, ranging from flour to cornstarch to arrowroot, each with its own unique properties. They are often added to food in the form of a roux, which is a starch and liquid mixture forming the basis for creamy pasta sauces, gravies, stir-fry sauces, and more. The main disadvantage of starch thickeners is that they need to be added in rather large quantities in order for substantial thickening to occur—a typical thick white sauce requires about ¼ cup of flour for every cup of milk. As a result, the food is diluted and often becomes much less flavorful.

The thickness of a sauce is often determined by a composite of the various ingredients, controlled by adding other ingredients than just starch. As an example, consider the sidebar on mac and cheese, which proposes an experiment for you to determine the effect of the ingredients (not only starch, but also cheese, butter, and milk) on the viscosity.

SIDEBAR 6: MAC AND CHEESE

Macaroni and Cheese

Ingredients
1 pound elbow macaroni
1 tablespoon olive oil
2 tablespoons unsalted butter
1 tablespoon all-purpose flour
1 cup whole milk
½ cup (115 g) American cheese
½ cup (115 g) Velveeta cheese

Directions
1. Bring a large pot of water to a boil over high heat. Add the pasta and cook until al dente, about 10 minutes, then drain. Add the olive oil to the drained pasta and mix well to keep the macaroni from sticking together.

2. Meanwhile, to make the cheese sauce, melt the butter in a saucepan over low heat.

3. Add the flour and whisk until there are no more clumps.

4. Slowly whisk in the milk.

5. Add the cheeses and cook, stirring, until the sauce is thick and smooth.

6. Remove from the heat and mix the cheese sauce into the pasta. ⚛

SIDEBAR 7:
VISCOSITY OF CHEESE SAUCE

In the mac and cheese recipe, all the ingredients come together to create a sauce with perfect viscosity; it's thick enough to cling to the pasta, but still thin enough that it flows. To determine the individual ingredients' effect on the viscosity, you can perform the following experiment. Prepare the cheese sauce as indicated in the recipe, then prepare three variations: one that omits the flour, one that omits the cheese, and one that omits the butter. You could even try preparing one sauce without milk, but this sauce will be so thick that it barely flows at all, so we suggest focusing on the other ingredients.

Directions

1. Fill a tall glass or graduated cylinder three-quarters of the way with cooked macaroni.

2. Once each cheese sauce is done, remove it from the heat and quickly measure the temperature (temperature affects viscosity so you want to make sure that the temperature is about the same for all four sauces).

3. Immediately pour the cheese sauce through a funnel into the glass with the macaroni, and start a timer. The funnel helps ensure that the sauce goes through the macaroni and not along the cylinder wall.

4. To calculate the speed of travel, measure how far the cheese sauce travels down the glass and note the time it takes to get there. For example, if the sauce travels all the way from the top of the glass to the bottom, measure the height of the glass and divide it by the time it took for the sauce to get there: this is your speed of travel.

5. Repeat with the other three sauce variations. Note that some sauces will reach the bottom of the cylinder and others will not. If the sauce cools down too much, you may reheat it gently until it reaches the temperature you noted.

Questions

When you are done, consider the following questions: Which sauce had the fastest speed of flow (lowest viscosity)? Which had the lowest speed of flow? Which ingredient is least important for the final viscosity of the sauce? Which is most important? How does temperature affect viscosity? Using the equation for volume fraction, how do the speeds of travel correspond to the volume fraction of each of the ingredients?

The photos in the figure show four versions of the mac and cheese sauce recipe, each with a different ingredient omitted. The classic recipe is sauce D, but variations without flour, cheese, and butter help us understand how each ingredient contributes to the

(A) No flour (B) No cheese
(C) No butter (D) All

final viscosity. Sauce B, which omits the cheese, has the lowest viscosity. It does not cling to the pasta at all and flows straight to the bottom of the jar. This tells us that cheese is the ingredient that contributes the most to the viscosity of the sauce. Omitting flour (sauce A) and omitting butter (sauce C) have similar effects on viscosity. In sauce D, which contains all ingredients, the flour, cheese, and butter combine to give a high-viscosity sauce that coats the pasta. Thank you to "karen_127" (one of the participants in our online class), who performed this experiment in her kitchen as part of a class assignment. ⚛

EMULSIONS

The third way to thicken a liquid food is to turn it into an emulsion. Since most foods are water based, this usually involves adding a fat such as oil or butter, and then breaking up one liquid into tiny droplets that float around in the other liquid. This creates a similar situation to using starch; droplets of one ingredient inside the other slow down the flow of the liquid. Many emulsion-based sauces, such as hollandaise and beurre blanc, involve mixing butter into water, wine, or vinegar to create droplets of butterfat in the other liquid. From a thickening perspective, the droplets work in a way similar to the swollen starch granules, although the underlying science is different. Since emulsions are such an important concept in food, and so difficult to control well, we will take a closer look at them in chapter 6. In the meantime, if you want to admire an emulsion, we suggest you make the following recipe for hollandaise sauce, from talented chef Nandu Jubany of the

Michelin-starred restaurant Can Jubany in Vic, Spain. It's fascinating how you can start out with two liquids, and by the time you are done, you have a thick solid-like sauce.

White Asparagus with Hollandaise Sauce au Gratin

Ingredients

16 fresh white asparagus spears, trimmed and peeled

2 large egg yolks

Juice of ½ lemon (20 mL)

160 g unsalted clarified butter*

Salt

Ground white pepper

15 g chives, finely chopped

Directions

1. Fill a large pot with salted water.

2. Tie the asparagus spears in a bundle, all pointing the same way. Place the bundle of asparagus in the pot with the tips pointing up; the water should almost cover them.

3. Bring the water to a boil and cook the asparagus until just tender. The cooking time will vary depending on the freshness, age, and origin of the asparagus, so it is important to adjust the time from as little as 4 minutes for very young asparagus up to 25 minutes for tougher spears.

4. Turn off the heat and let the asparagus stand in the cooking water.

5. Heat the clarified butter to 140°F (60°C).

6. In a small heatproof bowl, combine the egg yolks and lemon juice.

7. Place the bowl in a bain-marie and whisk until thick and creamy. When the temperature is 140°F (60°C), slowly start whisking in the warm butter. If the emulsion is close to breaking, whisk in 1–2 teaspoons of the asparagus cooking water.

8. When all of the butter has been added, taste and adjust for salt and white pepper, and add the chives. Keep the sauce in an 86°F (30°C) bain-marie.

9. Cut each asparagus spear into 4 pieces and place in a soup dish with a spoonful of the warm cooking water on top.

10. Add the hollandaise sauce on top and broil to brown lightly.

To make clarified butter, melt unsalted butter to 122°F (50°C), skim off the milk solids on top, and carefully separate the fat layer on the top (the clarified butter) from the water layer on the bottom. ⚛

POLYMER NETWORKS FROM FOOD COMPONENTS

Cooking often causes food components to leak polymers, which changes the texture of the food. We already saw that heating starch causes it to swell and clump together. But it does even more: as the starch granules are heated, they leak starch polymers in the form of long carbohydrate chains. These can entangle and form a network that further thickens the dish.

This basic principle occurs with many different foods. As an example, let's consider the amazing meat gravy recipe from chef Carme Ruscalleda. Carme starts her gravy with brined pork bones, which she browns in the oven. Partway through browning, she pours on some wine. After a while longer, she takes out the bones and adds some water to the pan and heats it on the stovetop to release all the fond (the burnt and stuck-on bits). She pours this mixture back over the bones and boils them together until all the flavor is extracted from the bones and the liquid has reduced significantly. Then she filters out the large pieces of bone and meat and does a final reduction on the stove until the sauce is thickened to the desired serving consistency.

This is a fantastic example of reduction. Each of the three reduction steps in the recipe serves a different purpose in terms of flavor, but in each case, the result is a slightly thicker liquid than when it started out. However, in addition to the reduction, there is a polymer at work in this recipe: gelatin. Gelatin is a protein polymer that is produced when the collagen in animal skin, bones, and joints breaks down. It forms a network—as you might expect by now—that contributes to the thickened sauce. Combined with particulate matter from the wine and bones, we get an altogether delicious and perfectly viscous sauce. The same principle underlies many meat stews. Of course, not all foods naturally contain gelatin or starch, which leads us to the fifth and final method for thickening sauces.

MODERNIST THICKENERS

What if your food has no inherent polymers, but you still want to thicken your food with them? Luckily for you, a large number of polymers have been isolated from plants and animals—and even microbes—just for this purpose. We often refer to these types of polymers as "modernist" thickeners, but their addition to food is nothing new—they have been added to foods in cultures around the world for millennia. They are fantastic for thickening because they are effective even when added in very small amounts. This solves many of the issues we encountered above. They do not dilute the flavor of the food the way starch-based thickeners do. They also do not involve the lengthy preparation times of thickening by reduction, nor the addition of large amounts of fat involved when thickening with emulsions. If you're a chef who wants to preserve flavor, modernist thickeners are a godsend.

Scientists often refer to these thickeners as hydrocolloids. This is really just a fancy word for particular types of polymers that enjoy being in water. Their name is a combination of *hydro*, referring to water, and *colloid*, referring to any mixture of well-dispersed but insoluble particles within another medium. Hydrocolloids usually come as proteins or carbohydrates. We've already encountered them in the form of the gelatin in Carme's meat gravy, and once you start looking for them, you'll find them in much of cooking.

How can the polymers thicken food without taking up more space? The short answer is that they spread out and fill large volumes, and they also entangle with other polymers. For a more detailed answer, let's examine what a polymer looks like in solution. A polymer is a very long molecule that is made up of many much smaller monomers. The bond between each of the monomers can bend in more or less any direction, making the polymer very flexible. Where have we encountered this movement in any direction before? That's right, in random walks. Indeed, it turns out that a polymer dissolved in a liquid will look very similar to the movement of such a molecule. You can imagine tracing the random walk of a calcium ion and ending up with a structure close to that of a polymer, where each step in the random walk corresponds to one of the monomers. In fact, the random walk of the polymer is even more spread out than the calcium ion because the calcium ion can move in the same spot several times, whereas the polymer is blocked by its own monomers. As a result, a long polymer will occupy a large volume simply by the way it arranges itself in the solution. Even more important, as the concentration of the polymer molecules increases, they no longer remain distinct and apart from one another; instead, they entangle, with one polymer molecule crossing through the space of many others. When they flow, they can no longer flow around one another, but are instead entangled and constrained. This is very similar to the alginate gel we already encountered, but here there are no permanent bonds, as was the case for the alginate. Nevertheless, the entanglements make it much more difficult to flow, and the viscosity can be increased significantly. This can occur even with the addition of small amounts of thickener.

Numerous hydrocolloids have been used in modernist cooking; gelatin is a common favorite, as it is naturally created from collagen falling apart when cooking bones. But this is just the tip of the iceberg. The sidebar shows how chef José Andrés, of the famed restaurant group that includes minibar and Jaleo, has utilized the power of several different hydrocolloids to create a remarkable dessert.

Almond Cake

Ingredients

Almond Cake Base (recipe follows)

Luxardo Cake Soak (recipe follows)

Almond Shells (recipe follows)

Almond Foam (recipe follows)

Black garlic purée

Marcona almonds, chopped and toasted

Grape Spheres (recipe follows)

Sherry Fluid Gel (recipe follows)

Directions

1. Brush the cake triangle with the Luxardo mixture.

2. Fill each shell halfway with almond foam, starting with the corners to form an even layer.

3. Pipe the black garlic purée in the shape of the letter M on top of the almond foam.

4. Fill the shell with more almond foam, leaving enough room for the cake (about 10 mm).

5. Sprinkle on a layer of toasted Marcona almond pieces, then top with the soaked cake triangle.

6. Invert the cake onto a chilled plate.

7. Garnish with three gold grape spheres.

8. Pipe 3 dots of sherry fluid gel to the right of the cake.

Almond Cake Base

Ingredients

160 g Marcona almonds
250 g egg whites
160 g egg yolks
120 g sugar
40 g all-purpose flour
Pinch kosher salt

Directions

1. Preheat the oven to 338°F (170°C). Line a baking sheet with parchment paper.

2. Spread out the almonds on the lined baking sheet and roast for 10 to 12 minutes, until golden brown and fragrant. Allow to cool completely.

3. In a Vitamix, blend the toasted almonds, egg whites and yolks, sugar, flour, and salt into a homogeneous, smooth batter.

4. Pass the mixture through a fine-mesh strainer and allow to rest overnight in the fridge.

5. Fill a 1 L siphon with 300 g of almond cake batter, charge with 2 nitrous oxide chargers, and shake vigorously.

6. Cut 4 small slits into the bottom edges of a pint-size deli cup (at 3, 6, 9, and 12 o'clock) to allow steam to escape while cooking.

7. Dispense 50 g of the almond cake batter into the vented deli cup.

8. Microwave for 56 seconds.

9. Allow to sit for 5 min to cool and set.

10. Unmold the cake from the deli cup and slice into a triangle (the shape of the base of the shell mold).

Luxardo Cake Soak

Ingredients
100 g Almond Milk (recipe follows)
50 g Luxardo Amaretto
0.75 g kosher salt

Directions
Combine the ingredients.

Almond Shells

Ingredients
0.8 g xanthan gum
600 g Almond Milk (recipe follows)
140 g heavy cream
3 g kosher salt
Liquid nitrogen

Directions
1. Line a rimmed baking sheet with a clean linen or dish towel and set in the freezer.

2. Using an immersion blender, blend the xanthan gum into the almond milk to thicken. Incorporate the heavy cream and season with the salt.

3. Pour liquid nitrogen into a shell mold to freeze it. Pour out the excess liquid nitrogen.

4. Fill the frozen mold completely with the almond shell base and allow it to set for 10 seconds.

5. Invert and pour out the excess almond shell base, creating a thin shell in the mold. Use a small palette knife to clean the bottom edges of the mold.

6. Splash the almond shell with a little more liquid nitrogen to help freeze, but be careful not to overfreeze and crack it.

7. Place the mold back in the freezer for a few minutes to ensure the almond shell is frozen.

8. Carefully remove the shell from the mold and transfer to the frozen linen-lined baking sheet.

Almond Foam

Ingredients
500 g Almond Milk (recipe follows)
2 silver gelatin sheets, bloomed (soaked in cold water for 5 to 10 minutes and drained)
2.5 g kosher salt

Directions
1. Heat 50 g of the almond milk to 158°F (70°C).

2. Remove from the heat and add the gelatin; stir to dissolve.

3. Thoroughly mix in the remaining 450 g almond milk.

4. Season with the salt. Store in the refrigerator.

Grape Spheres

Ingredients
150 g Grape Juice (recipe follows)
4.5 g calcium gluconolactate
0.5 g xanthan gum
Gold powder
Alginate Bath (recipe follows)

Directions
1. Weigh out 75 g of the grape juice and reserve the remainder for holding the spheres.

2. Use an immersion blender to blend in the calcium gluconolactate until completely dispersed.

3. Blend in the xanthan gum until thickened.

4. Blend in gold powder for color.

5. Allow the mixture to rest for 10 minutes, then blend again to homogenize.

6. Transfer the mixture to a container and use a vacuum machine to remove any air bubbles.

7. Carefully drop ¼-teaspoon spheres into the alginate bath; leave them there for 10 to 15 seconds to encapsulate. (The skins should be fairly thin.) Use a slotted spoon to carefully remove the spheres from the alginate bath and transfer to a bowl of clean filtered water to rinse off any excess alginate.

8. Transfer the rinsed grape spheres to the reserved grape juice.

Alginate Bath

Ingredients
5 g alginate
1 L filtered water

Directions
1. Use an immersion blender to blend the alginate into the filtered water until fully dispersed.

2. Strain the mixture through a fine-mesh sieve.

3. Allow the alginate bath to rest in the refrigerator overnight to dissipate any trapped air bubbles.

Sherry Fluid Gel

Ingredients
50 g Cepa Vieja sherry vinegar
0.3 g xanthan gum
200 g Almond Milk (recipe follows)
3.5 g kosher salt

Directions
1. Use an immersion blender to thicken the vinegar with the xanthan gum.

2. Slowly blend the thickened sherry vinegar into the almond milk.

3. Season with the salt.

4. Use a vacuum machine to remove any air bubbles.

5. Store the mixture in a squeeze bottle.

Grape Juice

Ingredients

500 g Himrod grapes, stemmed

0.2 g ascorbic acid

0.4 g malic acid

0.2 g kosher salt

Directions

1. Rinse the grapes in ice water to remove all dirt and sediment. Discard any dark or bruised grapes. Remove any black spots around the stem area. Spin through a salad spinner to remove all excess water.

2. Pass the grapes through a slow juicer into a container with 0.1 g of the ascorbic acid.

3. Strain the juice through a Superbag or fine linen cloth into a container with the remaining 0.1 g ascorbic acid.

4. Season with the malic acid and kosher salt.

Almond Milk

Ingredients

2 kg Marcona almonds

2 kg filtered water

Directions

1. Pulse the almonds and water in a Vitamix to break the almonds into smaller pieces.

2. Allow the mixture to rest overnight in the refrigerator to infuse.

3. The next day, blend the mixture again in the Vitamix.

4. Process the mixture through a slow juicer to extract the almond milk.

5. Pass the mixture through a Superbag or fine linen cloth.

6. Squeeze out the excess almond milk from the remaining pulp. ✾

Solids and Elasticity

Thus far we have limited our discussion to liquid foods. What about solids? A material turns into a solid once the volume fraction is high enough for it to become jammed. We already saw that regardless of the material, this happens at volume fractions of around 65%–70%. This number holds true whether you're using tomato chunks, starch granules, wine particles, a cup of flour, or any other particle-based cooking materials you may come across. Recall that once the tomato chunks were above this threshold, the marinara sauce could no longer flow, becoming a solid instead.

The essential property of a solid is that it has an internal structure that allows it to maintain its shape when pushed or pulled—to a point. For example, you can stand on a box and it will hold you up as long as you aren't too heavy; otherwise, you will break it. An ant, on the other hand, can stand on a piece of Jell-O or a steak and neither would break. We say that solid foods have *elasticity*. More specifically, a food's *elastic modulus* (aka Young's modulus) describes how stiff or squishy it is. The more difficult it is to deform a material, the harder it feels, and the higher its elastic modulus. Hard candy has a high elastic modulus and egg custard has a low elastic modulus.

Our mouths are extremely good at sensing elastic moduli, hence it is an important parameter to control when cooking. Figure 3 shows the different elastic moduli of some common foods.

Food	Elastic Moduli (kilopascal)
Bread	0.1–0.3
Jell-O	2
Banana (raw)	8–30
Peach (raw)	20–200
Potato (raw)	50–140
Apple (raw)	60–140
Carrot (raw)	200–400

FIGURE 3 The estimated elastic moduli of common foods are consistent with your experience: Jello-O is quite soft, while carrots are hard. Fresh peaches can be soft or hard depending on their ripeness. Elastic modulus is measured in Pascals, abbreviated Pa, which is a unit of pressure.

These numbers may look random, but they are not as esoteric as they may seem. In fact, you can find the elastic moduli of the foods all around you with a simple experiment. Simply cut a piece of food, about an inch long on each side. It can be a piece of cake, or steak, or whatever. Then place some weight on it, perhaps a small spice jar or anything else you can find. The food will compress by some small amount depending on how stiff or squishy it is. You can then calculate the elastic modulus by dividing the force pushing down from the heavy object by the area of the food it's pushing down on. (It's important to account for the area because the same force will compress a large piece of food less than a small piece of food.) Then, divide this number by how much the food has compressed in relation to how thick it was before compression. The result is your elastic modulus. When you're done, compare your numbers to the table and see how well you did.

If you think about it, this experiment is not a bad approximation of what actually happens to food in your mouth. When you bite down on a piece of steak, the force from your molar teeth pushes down on the steak and deforms it by some amount depending on how tough the steak is. At some point, the food snaps and breaks. We say that it fractures, or ruptures. Therefore, the elastic modulus is not just some random number—it comes pretty close to describing what occurs in your mouth as you chew your food.

SIDEBAR 10: MEASURING STEAK'S ELASTICITY

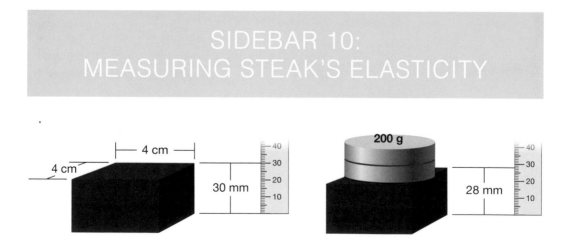

SCIENCE AND COOKING

A simple experiment to approximate the elastic modulus of a food is shown here with raw steak. Cut a square piece of steak, measure its sides, and calculate the area. In the image, each side of the steak is 4 cm, so the area is 4 cm × 4 cm, which is 16 cm², or 0.0016 m². Next, measure the height of the food before it is compressed (in the image we measured 30 mm); this is L0. Now, place a weight on top (it can be a spice jar or anything else you find in your kitchen, but ideally something that has a larger area than the piece of food). A perfect weight squishes the food only a little bit but doesn't completely smash it. Measure the new height of the food (in our case, 28 mm). Then ΔL (pronounced "delta L") is how much your food compressed, which in this case is 30 − 28 = 2 mm. As a last step we have to find the mass of the spice jar. Here, the total weight pushing down on the steak is 200 g. We have to multiply the weight in kilograms (0.2 kg) by 10 in order to convert the mass to a force, so we get a force of 2 N. Now, finally, you can plug your numbers into the equation for the elastic modulus!

$$E = \frac{\text{Force/Area}}{\Delta L/L_0}$$

$$E = \frac{(2 \text{ N }/0.0016 \text{ m}^2)}{(2 \text{ mm}/30 \text{ mm})} = 19{,}000 \text{ Pa}$$

You can try this experiment on steak before and after cooking and see how the cooking process changes the stiffness. Or try it on other foods and compare to the elastic moduli in the table. Are your values within the same range? If so, congratulations! You did the calculations right. ⚛

Ways to Increase Elasticity

There are countless ways to manipulate the elastic modulus when cooking, and a skilled cook is highly attuned to the delicate modifications required for even small changes in texture. Despite the complexities, the two main categories are the same as in liquid foods: you can increase the volume fraction or you can modify the polymer network.

MODIFICATIONS TO PACKING

By increasing the volume fraction, an already solid material can become even firmer. If you continue boiling off water from your marinara-sauce-turned-tomato-paste, the blob in your pan becomes stiffer and stiffer. If you spread out a thin layer of the paste in a dehydrator and evaporate even more water, your paste turns chewy and leathery. Its elastic modulus has gotten even higher.

This seems easy enough, and it is. It also applies to a large variety of foods. But even this idea has complications. Not only should we examine how *tightly* the chunks or molecules pack, but we should also look at *how* they pack. For some materials, this is what makes or breaks the dish. For example, when tempering chocolate, we must heat cocoa butter to a certain temperature and then cool it to another temperature in order to force the cocoa fat molecules to pack in a certain way. Only when packed in one of six possible crystal structures does the chocolate acquire the preferred sheen, snap, and long-term stability that we prize.

MODIFICATIONS TO PACKING AND POLYMER NETWORKS

As we now know, many foods can be manipulated via both packing and polymer network modifications. We've learned how cheeses like ricotta are made by first curdling the milk using rennet or acid to create a network of coagulated protein polymers. The next step is to strain or press the curds to remove the liquid whey. As whey is removed, the volume fraction of curds increases, and the cheese gets more solid. The final texture of the cheese depends on how much whey was removed. Tofu works in a similar way. Soy milk is curdled using salts like nigari or gypsum, then the curds are pressed until we achieve the desired level of firmness. Silken tofu, which is not pressed, feels like a soft custard, whereas firm tofu is much stiffer. In both cases, the final texture is a result of both the formation of a polymer network and the increased packing.

MODIFICATIONS TO GELS (POLYMER NETWORKS)

You may be surprised how many solid foods are actually some form of a gel. For gels, the elasticity comes from an interconnected network of components throughout the food that exists in water or air. The components can be shapes such as granules or long polymer chains, but what makes it a gel is that the individual units interact and are stuck together. When you push or pull at one edge, the parts farther inside will also feel the push or pull.

In order to truly understand elasticity in solid foods, we must spend some time discussing gels. Go ahead and imagine your favorite gel. We suggest an alginate gel from spherification or a cube of Jell-O, if you can't think of anything else. In your mind, zoom in to a microscopic view until you can clearly see the gelatin polymers and the crosslinks where they bind to each other at distinct points. Now imagine pulling on a crosslink until it gives way and the structure is disrupted. You have to add some energy to do this. If the gel is stiff, you must add a lot of energy; or, if the gel is soft, you don't have to add very much. This is the very idea of the elastic modulus on a microscopic scale. And we're in luck, because this idea is neatly summarized in the following equation:

$$\text{Elastic modulus} = \frac{\text{Energy}}{(\text{average crosslink distance})^3}$$

This equation holds the secret to manipulating texture in solid foods. It states that there is a certain energy needed to disrupt the very small volume surrounding each crosslink, conveyed by this expression: average crosslink distance^3. The ratio determines the elastic modulus.

This means that you could make a material firmer in two ways: by increasing the energy of the bonds or by increasing the number of crosslinks per volume. The most effective way is to change the average crosslink distance, because its effect is cubed, and this is usually what is done. To accomplish this, simply add more polymers, evaporate off water, or change the type of polymer. Chefs do all of these things, sometimes in combination. We will see how this nifty equation explains the textural changes in common, and not so common, foods.

GELS FROM HYDROCOLLOIDS

Jell-O is the quintessential gel. It is made of gelatin, water, and some sugar for flavor. From a scientific standpoint, gelatin gels are great because they contain only one sort of protein—gelatin—and as you increase the concentration, the gel gets firmer and firmer, behaving exactly as our equation predicts. Jell-O is also interesting in that it can transition back and forth between liquid and solid. The polymer chains stick to each other, but they do so rather weakly, so the gel solidifies when cold and melts again when heated. This is a very special property that is used to great effect in numerous dishes. One of them is Chinese soup dumplings (xiaolongbao), where the solid inside magically transforms into a rich soupy filling upon cooking.

SIDEBAR 11:
COREY LEE'S SOUP DUMPLINGS

Soup Dumplings

Ingredients
Dumpling Wrappers (recipe follows)
Lobster Filling (recipe follows)
Vinegar Sauce (recipe follows)

Directions
1. On each dumpling wrapper, place a ball of lobster filling.

2. Gather the dough around the filling by pinching the edges together in 18–20 folds.

3. Steam the dumplings in a bamboo steamer for 5 minutes.

4. Serve with the vinegar sauce for dipping.

Dumpling Wrappers

Ingredients
200 g all-purpose flour
195 g cake flour
5 g Starter Dough (recipe follows)
200 g water
2 g potassium carbonate and sodium bicarbonate solution

Directions
1. Combine the all-purpose flour and cake flour in a mixer.

2. Add the starter dough, water, and potassium solution and mix with a dough hook until evenly incorporated, approximately 5 minutes.

3. Feed the dough through an electric sheeter about 20 times, until smooth and elastic.

4. Wrap in plastic wrap and allow to rest at room temperature for 30 minutes.

5. Cut the dough sheets into 4.5 g pieces and form into thin circles, approximately 5 cm in diameter, using an electric sheeter or a small rolling pin.

Starter Dough

Ingredients
1 g instant yeast
50 g warm water
25 g cake flour
25 g all-purpose flour
0.3 g salt

Directions
1. Whisk the yeast into the warm water until completely dissolved.

2. Whisk in the cake flour, all-purpose flour, and salt.

3. Transfer to a mixer and knead with a dough hook for about 4 minutes, until a dough forms.

4. Remove from the bowl and knead with your hands until smooth.

5. Place in a bowl, cover, and leave in a warm spot to rise until doubled in size, about 2 hours.

Lobster Filling

Ingredients

360 g lobster stock
21 g gelatin, bloomed (soaked in cold water for 5 to 10 minutes and drained)
15 g soy essence
120 g lobster meat
10 g lobster coral, passed through a sieve
2 g salt
80 g clarified butter, whisked with liquid nitrogen until shattered like coffee grounds
16 g scallion, finely chopped
16 g fresh ginger, finely chopped

Directions

1. In a medium saucepan, bring the lobster stock to a boil. Whisk in the gelatin and soy essence. Allow to cool in the refrigerator until set, then chop finely. Measure out 380 g of the chopped gel.

2. Puree the lobster meat, coral, and salt until smooth. Add the clarified butter, scallion, and ginger.

3. Add the chopped gel and mix until evenly incorporated. Work quickly to prevent the gel and butter from melting.

4. Divide the mixture into pieces weighing 15.6 g–15.9 g and roll them into balls.

Vinegar Sauce

Ingredients
20 g water
20 g Banyuls vinegar
20 g black rice vinegar

Directions
Combine the ingredients and mix well. ⚛

Other gels do not behave like this. The alginate gels in spherification do not melt when heated, for example. Here, the calcium ions act as permanent bridges between the alginate polymers, and the bonds are strong enough to withstand heat. The result is a strong, heat-stable gel, which can be a desirable property if you want to serve your gels hot. Other polymer thickeners that behave like this include pectin (used in jams) and agar-agar (common in Asian cuisine).

As an example, here is a spectacular creation from Ferran Adrià that reimagines cod. He removes fish from its skin and replaces it with a gel. The gel combines the intense taste of codfish broth with black truffle. How can we make such a gel? Your first thought might be to use good old gelatin as the thickening agent, successful as it is in Jell-O. But this won't work. Gelatin behaves like a solid only when it is cold. To create a warm interior, we must use a gelling agent that does not fall apart at elevated temperatures. For this, Ferran uses agar-agar. This is a naturally occurring—and vegetarian—ingredient that stays a gel at higher temperatures and is the essence of the recipe.

Ferran's recipe replaces the critical ingredient of a dish with a hydrocolloid-inspired substitute that completely changes the experience of eating it. He has used intriguing ideas like this in many different dishes to create entirely new foods.

Warm Black Truffle Gel with Cod Skin

Ingredients
Warm Codfish Gel (recipe follows)
Black Truffle Gel (recipe follows)
Garlic Oil (recipe follows)
50 g pasteurized egg yolks
12 tender almonds

Directions
1. Cut the codfish gel in 1.5 cm by 8 cm rectangles.

2. Warm the dishes with the black truffle gel using a salamander broiler for 5–6 seconds. Place the codfish gel in the center of each dish and warm for an additional 10 seconds.

3. Brush the codfish skin with the garlic oil and place it on top of the warm codfish gel. Briefly warm again to temper the skin.

4. Add some egg yolk around the skin and place 3 almonds around it.

Warm Codfish Gel

Ingredients
4 (100 g) desalted codfish tails
400 g water
Salt
1 g powdered agar-agar

Directions

1. Scale the codfish tails, being careful not to break the skin.

2. Remove the skin from the meat. Preserve the skin between 2 sheets of wax paper.

3. In a pot, combine the water and the codfish meat. Bring to a boil and add salt to taste. Measure out 400 g of the codfish broth.

4. In a pot, mix the codfish broth and agar-agar. Bring to a boil over medium heat, stirring constantly.

5. Remove from the heat and briefly add air with an immersion blender to create a less dense gel. Transfer to a plastic container in which a thickness of 1 cm can be achieved and let set.

Black Truffle Gel

Ingredients
140 g black truffle water
Salt
0.35 g powdered agar-agar

Directions

1. Put the black truffle water in a small saucepan and season with salt to taste.

2. Mix in the agar-agar. Bring to a boil over medium heat, stirring constantly.

3. Remove from the heat and briefly add air with an immersion blender to create a less dense gel. Divide the mixture into 4 small deep dishes. Allow to set in the fridge for at least 2 hours.

Garlic Oil

Ingredients
2 garlic cloves
100 mL extra-virgin olive oil

Directions

1. Blanch the garlic cloves three times, starting from cold water each time. After the last blanching, refresh them in cold water and peel.

2. Cover the garlic cloves with the olive oil and cook at low heat for 2 hours. Leave the garlic in the oil for an additional 6 hours. ✵

An egg can serve as another example of a heat stable gel: once the gel of a solid egg has formed, there is no going back. And yet, even an egg holds surprises. Just as with packing, the *nature* of the polymer network will also affect the final texture of the food. To see what we mean, take a look at the sidebar for Carme Ruscalleda's Crema Catalana.

SIDEBAR 13: CARME RUSCALLEDA'S CREMA CATALANA

Crema Catalana

Ingredients

1 L milk
20 g thin lemon peel, coarsely chopped
3 g cinnamon stick
150 g sugar, plus more for the topping
45 g all-purpose flour
9 large egg yolks

Directions

1. In a small saucepan, bring the milk, lemon peel, and cinnamon stick to a boil, then remove from the heat. Allow the milk to completely cool down, about 1 hour.

2. Strain the milk, then divide the milk into two equal parts.

3. In a clean pot, combine half of the milk, the sugar, and the flour. Stir to make sure the sugar and flour are completely dissolved, then heat the mixture, stirring constantly, until it comes to a boil. Allow to boil for 15 seconds, then remove from the heat.

4. Add the remaining milk and mix well. Add the egg yolks and mix until homogeneous. Strain the mixture through a chinois into a clean pot.

5. Heat the mixture over medium heat, stirring constantly, until the temperature reaches 176°F (80°C). Strain again through a chinois.

6. Portion the cream into individual serving dishes. It is important to mix the cream while portioning, as this will give it more air, shine, and delicacy.

7. Use a blowtorch or traditional crema catalana sugar caramelizer (a round disk of iron with a wooden handle) to caramelize the top sugar layer. To heat the iron, press it against the stove burner or on the stove flame until it is flaming red.

8. Just before serving, sprinkle a generous layer of sugar over the custard and hold the blowtorch or hot iron closely over the top to caramelize the sugar. Serve immediately.

This crema catalana is the Spanish version of crème brûlée. The secret to the smooth silky texture of this dish is this: When eggs are heated quickly as in hard-boiled eggs, certain egg yolk proteins become scrambled by aggregating with each other. But in the environment of gentle heat, diluted with milk along with some sugar and flour to coat the proteins, the yolk proteins eventually form a fine network that gives the custard its elasticity. Crema catalana is an excellent example of how certain molecules can be coaxed toward a specific result by providing the correct conditions. ⚛

Cooking a Steak with the Thumb Test

From browning to heat transfer to protein denaturation, we've learned great lessons for how to cook the perfect steak. What about optimizing its texture? As you may suspect, a steak is also a gel. It is made up of a combination of cells, connective tissue like collagen (the basis for gelatin—which is why there is some debate as to whether Jell-O is a vegetarian dish or not, depending on what type of gelatin you use), and fat. A steak is much more complicated than Jell-O, but in essence it is still a gel. Like most gels, it consists mostly of water. Our bodies are 70% water, and cows are similar.

During the process of cooking a steak, the collagen contracts when it is heated past about 140°F (60°C). The contraction pushes water out and effectively concentrates the collagen. Additionally, the contracted fibers are stiffer than the uncooked ones. Together, these two effects combine to explain why a steak cooked to well-done is tough to chew. We have increased both the density of crosslinks and the energy of the bonds holding them together.

How might you test if your steak is cooked to your liking? It's likely you've heard of the famous thumb test. First, touch your thumb to your pinkie finger on the same hand. Use your pointer finger on the other hand to touch the muscle at the base of the thumb. The muscle is stiff—about the same stiffness as a well-done steak. Now do the same thing but touching your thumb to your pointer finger on the same hand. Your thumb muscle is now softer—about as soft as a rare steak. Touching your thumb to the other fingers will give you a range of steak donenesses between rare and well-done. Now go ahead and touch your steak and compare that perceived stiffness or squishiness to your thumb muscle. You can estimate the steaks' elastic moduli using nothing but your own hand!

Gluten and Bread

No doubt you have heard of gluten. Although much has been written about it as it relates to nutritional health and celiac disease, it is simply a molecule that is responsible for the elastic modulus in bread. Bread flour is 11%–13% protein, mostly in the forms of two proteins called gliadin and glutenin. When water is added, these proteins start to swell and form a gluten network.

Gluten is so central to bread's structure that it is extremely difficult to replace in gluten-free products. No other food molecule so effectively creates a strong elastic network. Gluten-free flours can roughly replicate a dough's appearance, but as soon as they are stretched, the lack of gluten becomes obvious. A normal gluten bread dough can stretch quite a bit before it breaks, but trying to stretch a gluten-free dough is pointless—it breaks apart immediately. When chef Mark Ladner, of the restaurant Del Posto in New York City, visited our class, he illustrated this by arranging a tug of war with the two doughs showing exactly this outcome. He now makes a gluten-free dough that uses a wheat flour substitute mixture with a small amount of xanthan gum. This long sugar acts as a binder to hold everything in place the way gluten would. Although it is not a perfect solution, it seems that this is the best we can do for now.

Plasticity

Let's close out this chapter with two comments on the texture of solid foods, namely the ideas of plasticity and brittleness. The concept of plasticity became apparent in the previous section on bread dough, which is an interesting material because it will not spring back when you pull on it. In contrast, finished baked bread does spring back, at least as long as you don't push too hard. So does steak, cake, candy, or any of the other things we have previously examined. We talked about these spring-like solid foods as having elasticity. But when a food does *not* spring back, we instead say that it has plasticity. Bread dough is plastic. Plasticity is still governed by the same principles of packing and polymer networks as other solids, but now the polymers or particles that make up the food can very slowly slide past each other, forming new bonds so that they stay that way.

For some kinds of cooking, this is just what you want. If you've ever made strudel, the European pastry in which a very thin layer of dough is rolled up around a sweet fruit-based filling, you have experienced this. In order to achieve the thin, transparent layers, the dough is slowly and carefully stretched until it is very thin. To achieve this, the gluten network must be sufficiently strong; otherwise, the dough will crack and fall apart while it is being stretched and rolled. Here is a strudel recipe made by our friend Bill Yosses, the former White House pastry chef. Strudel is a fascinating food: by working the gluten, the dough becomes remarkably elastic. Like pizza dough, the strudel dough can be stretched into a large, thin sheet without rupturing. For strudel this makes it possible to roll up the delicious filling. We cannot do this with bread dough. The critical difference for strudel (and pizza) dough is the oil in the dough recipe, which modifies the interactions between the gluten strands.

SIDEBAR 14: BILL YOSSES'S STRUDEL

Strudel

Ingredients
12 ounce high-gluten or bread flour
1 teaspoon salt
2 tablespoons vegetable oil
1 cup warm water
12 apples, peeled, cored, and diced small
1 cup bread crumbs
1 cup sugar, plus more for sprinkling
1 cup golden raisins, plumped in hot water overnight, drained, and dried on paper towels
1 teaspoon ground cinnamon
Clarified butter

Directions
1. In a mixer, combine the flour, salt, and oil. With the mixer running, slowly pour in the water. Let the mixer run for 10 minutes, or until an elastic dough is formed.

2. Rest the dough for at least 2 hours.

3. Preheat the oven to 350°F (177°C).

4. In a large bowl, toss together the apples, bread crumbs, sugar, raisins, and cinnamon and allow to drain in colander.

5. Place a 4 ft square cloth on the table and sprinkle lightly with flour.

6. Stretch the strudel dough carefully and lay it on the cloth. Brush it with clarified butter and sprinkle lightly with sugar.

7. Cut off the thicker ends of the dough to leave only the thinnest layer of the stretched dough.

8. Spread the filling in a 3-inch-wide strip horizontally on the part of the dough closest to you.

9. Use the cloth to lift the edge of the dough and bring it over the filling to tuck in the other side. Continue to raise the edge of the dough closest to you and roll the strudel dough over on itself to roll into a cigar shape cylinder.

10. Bake for 45 minutes. ✺

Brittleness

We know that packing and polymer networks form the basis of both viscosity and elasticity. But there are other mouthfeels that can be derived from them, too. As one final example, let's return to the marinara sauce with which we started this chapter. When we last left it, we had evaporated enough water with the help of a dehydrator so that it was chewy and leathery. Imagine now that we had spread the paste really thin and continued evaporating water so that we managed to get rid of almost all of it. The result would be a new kind of texture, like a potato chip made of tomato, if you will. If you pushed on this chip, it would almost instantly break. We call these types of materials "brittle." Think of hard praline, chocolate, or crackers. It's a popular texture in the culinary world because it makes for such interesting mouthfeels. Note that you've achieved it by manipulating the same principle as before—by evaporating water and increasing the volume fraction—but now, with all the water gone, there are fewer bonds to hold the material together and it breaks very easily.

We hope to have shown you in this chapter how manipulating only two parameters can capture a majority of the textures and mouthfeels we experience in foods. You can now think of the many words you might use to describe food textures and test them against these principles. Some will fit, others may not—the multitude of ways in which we experience food in our mouths is vast, and we do not profess to have covered all of them. Nevertheless, we have seen yet again how a few basic scientific principles can illuminate what goes on in our food. Ponder this the next time you take a bite of your favorite food.

CHAPTER 6

Emulsions and Foams

Emulsions and foams structure flavorful liquids so that they flow slowly, they cling to foods, they cling to our tongues and our palates, and prolong our pleasure.

—Harold McGee

Mixing Ingredients That Don't Mix

Part of the fun and challenge of cooking is mixing ingredients together. Sometimes mixing is delightfully simple: a cup of water can absorb a *lot* of sugar, and you don't have to work very hard to make it dissolve. Other ingredient combinations are more difficult. For example, you might want to make a vinaigrette for your salad, but the two main ingredients—oil and vinegar—don't naturally mix. Vinegar is mostly water, so no matter how hard you shake a bottle of oil and vinegar, if you let the mixture sit for a little while, the oil and vinegar will separate. You can pour the dressing onto the salad and eat quickly, or you might decide to make the mixture in an electric mixer instead, at high speed. Surely that will work. Alas, even with high speeds, you will notice that the vinegar and oil still separate, though it might take a bit longer. One day, however, something magical occurs: you decide to spice up your salad dressing by adding a bit of mustard to the oil and vinegar. Now when you shake the dressing, the mixture stays together much better. Why is this?

The science should be familiar by now: Water prefers to stick with other water molecules—so much so that they try to get as far from the oil molecules as possible. Similarly, oil molecules prefer to stick with other oil molecules. If oil and water are placed in the same container, the molecules are happiest when they form two separate layers of liquid: oil on the top (since it is less dense) and water on the bottom. Fluids like these that cannot dissolve in one another are called *immiscible*.

Because oil is hydrophobic, it is immiscible in water. It is impossible to mix water and oil at a molecular scale. Instead, they form drops of one fluid (the oil) in the second (the water). But the mixture is still unstable because there are now so many interfaces where oil and water meet. So, as the oil drops rise to the top, they touch one another and merge. Ultimately, the two fluids completely separate, leaving just a single interface between them. So then how does mustard help? The answer is that the mustard forms a buffer between the oil and the water, and this prevents two droplets from merging to form a single one. This stabilizes the drops and lets you eat the vinaigrette whenever you want. The stabilization works because the molecules that make up mustard are amphiphilic: one half of each molecule is hydrophilic and likes to be in water, whereas the other half is hydrophobic and likes to be in oil. Amphiphilic molecules, such as mustard, are called surfactants, or surface active molecules. Here's why this solves our problem: When we mix oil and vinegar to make a vinaigrette, we form drops of oil in vinegar. The harder we shake, the tinier the droplets get. If we zoom in and look at the very edge of any drop, we see a miniature version of the separated layers that we would get without mixing. The mustard molecules sit at this interface. They are oriented so that each end is touching its preferred liquid, either oil or water, acting as a sort of intermediary that ultimately stabilizes the droplets. Thus, the ingredients appear mixed to our eyes, even though at a microscopic level they are not truly mixed. We call this state of matter an *emulsion*, where there are droplets of one fluid inside a second fluid. The droplets are small enough that the fluids feel mixed when we eat them. In fact, the mixture of the two fluids that form the emulsion provide a different, and often very enjoyable, texture and taste. In this chapter, we will focus on how emulsions work: How do we make them? How do they form? And, of course, how can we use emulsions to make truly amazing foods?

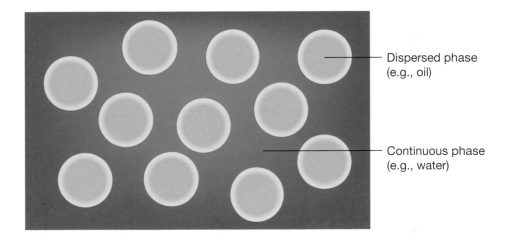

Dispersed phase
(e.g., oil)

Continuous phase
(e.g., water)

FIGURE 1 An emulsion is made up of drops of one liquid (the dispersed phase) in another liquid (the continuous phase). When the dispersed phase is oil, we call it an oil-in-water emulsion. When the dispersed phase is water, we call it a water-in-oil emulsion.

Droplets, Droplets, Everywhere

Examples of emulsions in foods abound: salad dressing, hollandaise sauce, and mayonnaise are just a few of the more straightforward examples. Each of these is a mixture formed from two fluids, the texture of which is dramatically different from how it would feel if we consumed the ingredients separately. We refer to the fluid inside the drops as the dispersed phase, while the fluid surrounding the drops is the continuous phase. Even foods like cookie dough and meatloaf are emulsions, although the presence of the two fluids is less obvious; in these cases, the dispersed phase is still a fluid, while the continuous phase is more of a gel or a solid.

But let's return to the simple vinaigrette with olive oil and vinegar. If you happen to have some oil and vinegar handy, feel free to pour some of each into a bottle with a tight-fitting lid and follow along. The classic thing to do is shake the bottle before pouring the vinaigrette onto your salad. This creates tiny drops of oil within the vinegar. You can tell that an emulsion has formed because the mixture is no longer transparent—small drops do not let light pass through the way it would in pure water or oil, so the mixture appears opaque. To make an emulsion, we can use anything that allows us to mix the two phases: a bottle to shake, a whisk, a blender, or a mortar and pestle can all work. The important thing is to form drops of one phase within the other. A gentle shake creates large drops, while a more vigorous shake for a longer time creates much smaller drops. In both cases, however, the oil drops will eventually float to the top, merge, and form a separate layer; large drops will separate quite quickly, whereas smaller drops will take a bit longer.

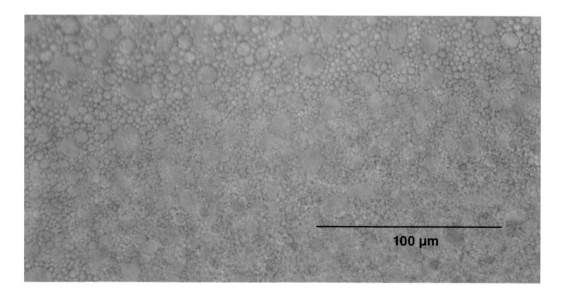

FIGURE 2 Although mayonnaise looks opaque to the eye, the microscopic oil drops can be seen with a microscope. The mayonnaise in the picture has oil drops that are anywhere from 2 to 20 micrometers in diameter, less than a tenth or half the width of an average human hair.

Making drops is one thing, but keeping them dispersed is another thing entirely. As our experience with oil and vinegar teaches us, an emulsion can be very delicate; the wrong kind of agitation or the wrong kind of conditions can cause the emulsion to break and the two fluids to separate.

Let's look at some real-world examples to illustrate why emulsions separate. The two liquids really dislike each other, so we can think of them like the Red Sox versus the Yankees. Dogs versus Cats. Real Madrid versus FC Barcelona. Just as fans of opposing teams sit in different sections in a stadium, the molecules in the different liquids tend to collect in different parts of the bottle, one on top and one on the bottom. By doing this they minimize the surface area where the molecules have to touch each other. Think of it like this: If every Red Sox fan sits next to a Yankees fan, there will be fights and disagreements everywhere. But if we put all the Red Sox fans in one section and all the Yankees fans in another, we run the risk of having arguments only along the border between the two sections. We have minimized the surface area where they have to interact. Even if we can't have a perfect division into two halves, we're still better off if we can create several smaller subsections of opposing fans. The surface area of interaction is still smaller than if they are completely mixed.

The same idea applies to drops in an emulsion. If you mix the emulsion really hard to make many small droplets, then if nothing is stopping them, the droplets will merge with each other, making larger droplets. Pockets of Red Sox fans in a stadium prefer to celebrate together and stay away from the dreaded Yankees, and vice versa. When this happens in an emulsion of oil droplets in water, the lighter oil drops will float to the top and the heavier water will sink to the bottom. Along the way, if two drops collide with one another, they can combine to create a single larger droplet—this is called *coalescence*. Coalescence can be a very fast process; for example, two drops of oil in vinegar may take only a few thousandths of a second to merge. Even if the drops don't collide with others as they rise in the mixture, once they reach the top, all the drops of oil will begin to touch their neighbors as the water sinks to the bottom. Thus, the process of coalescence repeats over and over until the two liquids have completely separated.

FIGURE 3 Olive oil and balsamic vinegar naturally separate into two layers, with the oil on top and vinegar on bottom (top left). When the mixture is shaken, the two layers mix and create a homogeneous solution (top right). Relatively quickly, though, the two fluids start to separate again; the photo on the bottom left shows the dressing 20 minutes after shaking, when the smaller drops have coalesced into larger drops that are visible to the naked eye. Finally, in the photo on the bottom right, taken 1 hour after shaking, the oil and vinegar have almost completely separated again.

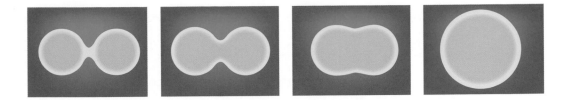

FIGURE 4 Emulsions separate into two layers because of coalescence of individual droplets into larger droplets. When individual droplets are far apart, they are distinctly spherical. If they bump into each other, the molecules of two droplets start to form a connecting bridge, which eventually leads to coalescence. The temporary dumbbell shape is not stable, but the molecules endure it briefly in order to merge into a larger droplet that has a smaller total surface area than the two smaller ones.

How can we prevent these mergers of droplets? Continuing the sports analogy, we cannot make Red Sox fans like Yankees fans. It is just impossible. The only way to keep the Red Sox and Yankees fans from seeking out their own kind is to make some physical barrier between them so that they are not so upset about being seated next to each other. Similarly, we cannot make water and oil like each other, but we can build a molecular divider to stop the droplets from merging. Imagine a person who loves baseball but doesn't particularly care more for either the Red Sox or the Yankees. Perhaps they grew up in Alaska and are now on vacation in Boston and New York. They might have some affection for both teams and will be equally happy sitting next to either sort of fan. And, importantly, if they are seated between two opposing fans, their presence will prevent the two fans from fighting with each other. In fact, the fans on either side might prefer to have this buffer between them, and thus be less likely to go try to find their friends.

The molecular equivalent of such a person is called an *emulsifier* or *surfactant*. The surfactants sit at the interface between the two phases. They coat the droplets and, by doing so, shield them from each other and stop them from merging. Mustard seed hulls contain molecules with these properties, which is why mustard stabilizes vinaigrette. In contrast, salt and pepper are not surfactants, so they have no effect on emulsion stability. For a more thorough discussion on surfactants, see the sidebar.

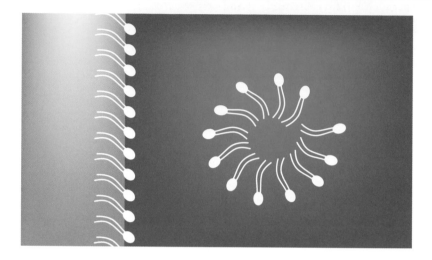

Surfactants sit at the interface between oil and water with a fixed orientation: the hydrophobic "tails" are in oil and the hydrophilic "heads" are in water. Their primary purpose is to coat the surface of the oil and water droplets and hence increase stability. If there is too much surfactant in the emulsion and there are no more interfaces left for them to coat, the excess molecules will form tiny droplets of their own. These are called micelles, and they can be pointed in either direction depending on whether the continuous phase is oil or water. In the image, the hydrophobic oil-like tails are pointing toward the micelles' centers to prevent them from being in contact with the water. Surfactant molecules constantly exchange between the droplet interface and the micelles, but the total number of molecules on the interface at any point remains the same.

Surfactants vary in shape, and this can make them more or less suitable for different types of emulsions. For example, if the water-loving heads are wider than the tails, the surface will naturally curve as the heads crowd the surface. This type of surfactant naturally favors oil-in-water emulsions where the surfactants' heads will be on the outside of the droplets. Conversely, if the tail is wider than the head, the surface will naturally curve with the tails outside, and this will be preferable for water-in-oil emulsions where the tails orient themselves on the outside of the drops. ⚛

Types of Surfactants

You might be surprised to learn that molecular mediators are quite common, though the mechanisms through which they work are diverse. Surfactants naturally exist in all kinds of food ingredients—garlic, mustard, eggs, starch, and more. Broadly speaking, the reason there are naturally occurring surfactants is because many of the molecules in biology have reasons to be both hydrophobic and hydrophilic. Recall, for example, our discussion in chapter 3 on the role of these opposing affinities in the folding of proteins. To highlight the diverse ingredients that act as stabilizers, let's explore the different ways that they work.

PHOSPHOLIPIDS

Phospholipids are the molecules that form the cell membranes in our bodies. Cell membranes need to separate the inside of the cell from the outside, and they do this by having molecules with a hydrophilic head and a hydrophobic tail. In a cellular environment, these molecules make membranes; they form a bilayer structure where hydrophobic parts of the molecules orient themselves to face the hydrophobic parts of other molecules. It is a bit like Velcro, where each of the adhering sheets represents a single layer of molecules that adheres to another layer that is oriented the opposite way. The same molecules that form membranes in cells can also act as surfactants when added to oil and water mixtures. One common surfactant that works this way is lecithin, which is found in egg yolks and various plants like soy. By dissolving some soy lecithin into the vinegar before mixing in the oil, we can make a more stable vinaigrette.

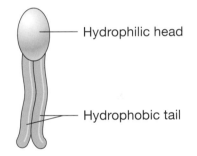

Hydrophilic head

Hydrophobic tail

FIGURE 5 The structure of lecithin, a commonly used surfactant for home cooks and in haute cuisine. Like all surfactants, lecithin has the critical feature of having a head that is hydrophilic—it likes being in water—and a tail that is hydrophobic—it hates being in water and likes being in oil. These properties allow it to sit happily at the interface of the droplets in an emulsion.

SMALL SOLID PARTICLES AND PROTEINS

Other than amphiphilic molecules, there are two other ways to stabilize the oil-water interface. The first is small solid particles. Just like molecular surfactants, particle stabilizers sit at the oil-water interface. They are able to do this because, just like our Alaskan baseball fan, the particles are literally neutral—they don't mind touching either fluid and can touch both at the same time. Since particle stabilizers are much larger than molecular surfactants, they are much harder to pull from the interface of the drops and are thus very effective stabilizers, in general much better than molecular surfactants. Examples within cooking include starch granules, polysaccharides like pectin, fat globules, and even yeast cells.

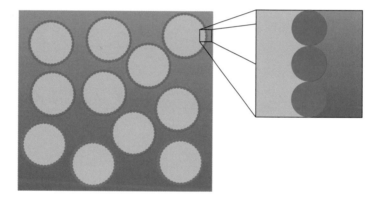

FIGURE 6 Particles (green) are much larger than individual oil and water molecules, so they can serve as a special type of armor for the droplets. The particles must coat the entire surface of the drops to be effective stabilizers, as otherwise their fluid can still touch and coalesce when drops collide.

Proteins offer another way to stabilize the oil-water interface. Recall that proteins have parts that are both hydrophobic and hydrophilic, and that a folded protein works by hiding its hydrophobic parts from water. Proteins thus have the ability to act like both particles and molecular surfactants: when they are folded, they act like particles; when they unfold, the hydrophobic and hydrophilic parts can separate and orient themselves according to their preferences at the interface.

Other Ways of Stabilizing Emulsions

There is also a way to stabilize emulsions that does not rely on particles or molecules at the oil-water interface. This method aims to prevent coalescence of droplets, which, as we learned, is one of the main reasons emulsions fail. Coalescence requires the droplets to come into contact with one another, so if we can prevent the collisions, we should also be able to stabilize the emulsion. All it takes is transforming the continuous phase into a medium that does not allow collisions.

There are three options for doing this. First, we can make the continuous phase, which is originally fluid, into a gel, as seen in Figure 7. This would prevent the droplets from colliding into each other. Second, we can change the continuous phase into a solid, either with some other type of phase transition or by achieving some type of jamming threshold. Third, we can make the continuous phase into a highly viscous fluid. This last method isn't quite as good as the other two, since even if the medium is very viscous, the droplets have motion and will eventually collide. All of these effects—gelation, solidification, and viscosity—can be controlled with temperature, and for that reason changing temperature is often a useful way to stabilize an emulsion in cooking. You will notice, for example, that many emulsions are stored in the refrigerator and consumed cold.

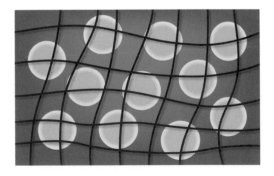

FIGURE 7 Gels stabilize emulsions by preventing the drops from colliding and coalescing. The gel can be made with a network of hydrocolloids, which holds the droplets in place. Although the droplets are often too small to be seen with the naked eye, they are much larger than the hydrocolloid polymers, which are flexible and can bend around the drops. In order to move, the drops have to push the polymers aside. Depending on the type and amount of thickener, the continuous phase can become very viscous, or even solid, and this tends to make the overall texture of the emulsion thicker.

With these scientific principles in mind, let's explore some recipes for sauces and desserts where these principles are at play.

HOLLANDAISE

Hollandaise, an emulsion of butterfat in water, is a classic French sauce with a reputation for being tricky to make. In chapter 5, we learned how to make the classic recipe by Nandu Jubany. Hollandaise must be prepared as a warm emulsion because the butter would be solid at room temperature.

The sauce is stabilized with egg yolks, which contain lecithin, a natural surfactant. As explained earlier, this surfactant coats the fat droplets and shields them from merging. But the egg yolk also performs a second function: its proteins unfold upon gentle heating and when making contact with ions in the lemon juice, form an entangled mixture of long, sparse protein polymers. This makes the water phase more viscous and also helps stabilize the emulsion.

To do this correctly, you must slowly drizzle in the melted butter while whisking to ensure that there is only a small amount of butter at any time that can create the tiny drops. As the butter is added, the volume fraction of the butter increases, which makes the emulsion more viscous. But this same effect also makes it harder to keep the emulsion in one piece. The increased volume fraction makes the droplets collide much more often.

Furthermore, there is a lurking disaster: if too much of the butter is added at once, the two phases spontaneously switch places in a process called phase inversion, such that we end up with water droplets in a sea of oil. Oddly enough, such a sauce, in which oil drops are dispersed in water, tastes and feels very differently from the intended recipe, even if the ratio of ingredients doesn't change. A fat-in-water hollandaise tastes creamy and smooth, but a water-in-fat hollandaise feels greasy in the mouth. Emulsions naturally want to invert when the volume fraction of the dispersed phase (in this case the butter) surpasses 50%; an inverted emulsion means that the new dispersed phase will now have the lower volume fraction. Preventing inversion is difficult. When the volume fraction gets high, you must pay careful attention to the emulsion as you add more ingredients and be certain to stop in time.

Other disasters also lurk: for the recipe to work, you must ensure that all ingredients are at the same temperature when added to the emulsion. Notice that the recipe instructs you to keep the asparagus in its warm cooking water, and that the butter and egg yolks are heated separately. If you fail to follow these steps, the emulsion will break. For example, butter will solidify below 35°C (95°F). Working at 60°C (140°F) ensures that the fat remains above that temperature. The higher temperature also decreases the viscosity of butter so that the drops can more easily be broken up into smaller drops.

The laws of physics conspire so that hollandaise sauce is not easy to make. As with so many things in life, it takes practice.

SIDEBAR 2: MAYONNAISE

Mayonnaise

Ingredients
2 large egg yolks, at room temperature
1 teaspoon mustard
½ teaspoon salt
2 cups (475 mL) olive oil
1 tablespoon (15 mL) lemon juice

Directions
1. Combine the egg yolks, mustard, and salt in a medium bowl.

2. Very slowly pour in 1 cup of the oil, drop by drop, while whisking vigorously. The mixture should thicken slightly.

3. Whisk in the lemon juice. Very slowly stream in the remaining cup oil, still whisking vigorously. If at any point the oil remains visible, stop drizzling and just whisk. Once the mixture is smooth and thick, cover and refrigerate until ready to use.

4. If the mayonnaise breaks, put a teaspoon of water in a clean bowl and start the process over by slowly adding the broken mayonnaise while whisking. ❀

One of the most remarkable properties of emulsions is that they can also be a solid. Let's compare the sidebar recipe for mayonnaise, a solid emulsion, to Nandu Jubany's hollandaise sauce in chapter 5, a fluid emulsion. A quick look at the ratios of ingredients shows a key difference: the amount of water (in the form of lemon juice) and number of egg yolks are similar, but the hollandaise contains about three-quarters of a cup of clarified butter (160 g) compared to the mayonnaise's 2 cups of oil. This means that the hollandaise has a fat volume fraction of about 83% and the mayonnaise has a fat volume fraction close to 95%. (Both of these calculations assume that an egg yolk is half water and half fat.) That is a lot of fat for both emulsions, but especially so for the mayonnaise. It will have many more oil drops packed into it than the hollandaise sauce. It will be a stiff solid emulsion. Looking at the volume fraction alone, we would expect the hollandaise to also be a solid—recall that the jamming threshold is 65%–70%. But hollandaise sauce flows easily, so clearly it is not a solid. There are two aspects to the recipe that could help explain why: First, hollandaise is served warm, so the butter fat is fluid. This not only decreases the overall viscosity of the sauce but also contributes to the instability of the emulsion—one of the secrets is to keep both the eggs and butter at the same temperature when whisking them together. Second, the mayonnaise contains extra emulsifier in the form of mustard. With a volume fraction as high as 95%, the oil drops in the mayonnaise are squished together to the point where it's impossible for the droplets to remain spheres. They must deform. At the same time, the droplets would prefer to remain as spheres, since this shape minimizes the area of contact between the water and the oil. Deforming the droplets is a lot like squishing a balloon; they yield slightly but can easily break and release their contents. The surfactant in the mustard helps prevent this from happening. Still, both emulsions are very delicate and can easily fail. Or as Harold McGee once put it in our class: "Olive oil mayos are like time bombs. They're slowly ticking away toward their ultimate destruction."

GARLIC AIOLI

Aioli is a traditional sauce from the Mediterranean region made with garlic and olive oil. In Catalonia, where Nandu's Michelin-starred restaurant is located, the sauce is known as *alioli*, which translates directly to "garlic and oil." Many restaurants and chefs nowadays use egg yolks to stabilize their aioli because it is much easier—eggs have a lot of surfactant, making it pretty straightforward even for a novice to manage to coat the droplets of the dispersed phase with the surfactant. But the original sauce uses only garlic, oil, and salt. These appear to be strange ingredients for an emulsion. Where are the surfactants? Where is the water?

It turns out that garlic itself contains a small amount of water, in which the oil can form drops. It also contains a small amount of emulsifier, which is released when the garlic is mashed up. It is extremely tricky to make an emulsion with so little water and emulsifier, but this is how you do it: You first grind the garlic into a paste, and then add coarse salt to help break the cell walls and draw out the water. Then oil is added drop by drop, very slowly and with lots of whisking. Making an aioli this way takes much patience and time, and every time the emulsion starts to break, you must add a few drops of water. The volume fraction of oil in the final product is so high that adding too much oil at once toward the end could destroy all your hard work. The water droplets help maintain the continuous phase. Experienced chefs listen for a characteristic "aioli sound" to determine when the sauce is ready.

Garlic Alioli

Ingredients

25 g peeled garlic cloves
5 g coarse salt
450 g refined (0.4°) olive oil
25 g water, as needed

Directions

1. Place the peeled garlic and salt in a mortar. Using the pestle, crush the garlic mix into a fine paste.

2. Add the olive oil little by little, making sure to stir thoroughly and constantly to emulsify with the pestle. It is important to not stop stirring. If the mix thickens too much, warm drops of water can be added while continuing the emulsification.

3. The alioli is done when the pestle can stand up straight in the emulsion. ❀

CHRISTINA TOSI'S AMAZING COOKIES

Who doesn't love a rich, decadent cookie? Recall Christina Tosi's cookies from chapter 1 of this book. She makes her cookies with far more butter and sugar than most cookies we encounter. For the same amount of flour, she uses 1.5 times as much butter and 2 times as much sugar as in a standard batch of Nestlé Toll House cookies. The creaming method typically used for Toll House cookies would never be able to incorporate so much butter and sugar without the batter feeling oily and gritty. Christina accomplishes this impressive feat by increasing the total mixing time to 10–15 minutes in a stand mixer, which results in a well-emulsified cookie dough. She also adds the egg yolks cold so that they offset the heat caused by the mixing, which helps keep the emulsion at an ideal temperature. Because the butter and sugar are creamed in the first step, some air is incorporated as small bubbles into the dough as well; bubbles of air in a continuous phase form a foam, meaning that the dough is both a foam and an emulsion.

BUTTER

Butter is itself an emulsion, but it is made by exploiting the phase inversion process that we try so hard to avoid in other recipes. Butter is made from cream, which is an emulsion of fat globules suspended in water. When cream is shaken past the whipped cream stage, the fat eventually clumps together and creates a large mass of butterfat, leaving behind liquid buttermilk. We can make butter from this by washing off the butterfat to remove the sugars and proteins that shorten its shelf life. In its final form, butter contains 15%–30% water in the form of small drops within the fat.

Foams

Here is a simple experiment we like to use in our class. We put a few egg whites in a small bowl and an equivalent amount of water in a second small bowl. We then ask for two volunteers to come up. Without telling the class what's in the bowls—from a distance the two transparent liquids look the same—we tell the volunteers to whisk the material in their bowls as rapidly as possible. The person who is lucky enough to get the egg whites ends up with a beautiful foam. The person with the water gets only water. The experiment shows that even though the initial liquid looks much the same, what you end up with is remarkably different.

Foams are similar to emulsions, except they have air dispersed as bubbles in a liquid, instead of drops of one immiscible liquid dispersed inside another liquid. They are, if anything, even more remarkable than emulsions. With foams, you can mix a liquid and a gas, and voilà—you have a solid! In the experiment in our class, the foam is a solid made out of the liquid and air; the water remains a liquid. During the whisking process, you are pulling a lot of air bubbles into both the egg whites and the water. But, unlike the water, the egg whites have something that allows the bubbles to remain incorporated and stable.

Of course, we know what that something is: protein emulsifiers that protect the little bubbles we create through whisking. In egg white foams, the emulsifiers are unfolded ovalbumin proteins. But because the emulsifier is a protein, it will be affected by pH. If pure egg whites are whipped, they will initially form a very nice foam, but the liquid will eventually drain out. By adding a touch of acid, such as lemon juice or cream of tartar, the foam will be much more stable. As we learned before, this is because adding acid denatures the proteins a bit, thus exposing more hydrophobic parts. In the case of foams, it is the air that plays the role of the oil, and the hydrophobic parts of the surfactant naturally go to the air instead of the water.

Haute cuisine chefs have been interested in creating unusual foams since the invention of the culinary foam by Ferran Adrià. He had a brilliant idea: Suppose you want to make a foam with some special flavor—for example, raspberry. You might choose to create the foam by adding egg white to the raspberry juice and whisking. This will make a stable foam. Unfortunately, however, it will taste like egg. This is because the egg has strong flavor molecules that overwhelm the taste of the raspberries. How can we make a foam out of raspberries that tastes like raspberries? Ferran's great idea is to add a natural surfactant to the raspberries and use this to allow the creation of the foam instead.

This idea led to a revolution in cooking. Chefs often call culinary foams simply "airs," referring to their light, airy texture. The menu at elBulli featured foams made out of an amazing range of ingredients, from savory potato and basil foams to sweet coconut foams as part of dessert. Likewise, Joan Roca makes foams of

Comté cheese and Chablis, and Nandu Jubany makes a delicious foam using sea water, which he serves on top of an oyster in a shell (see the sidebar for the recipe). Among more everyday foods, anything that feels spongy or light for its size is probably some type of foam. Whipped cream, the crema on top of an espresso, and the head on the top of a well-poured beer are all foams that we encounter daily.

Oyster Foam

Ingredients

20 g oyster water (reserved from when the oysters were opened)

0.1 g soy lecithin (this is the surfactant that makes this recipe work)

Directions

1. In a small saucepan, mix the oyster water and lecithin.

2. Carefully heat the mixture, making sure it doesn't exceed 104°F (40°C).

3. Using an immersion blender, blend until a layer of foam has formed on the surface. Remove with a spoon and place on top of the opened oysters. ⚛

Foams are close cousins of emulsions, except that the dispersed liquid phase is replaced by a gas phase. Thus, instead of liquid droplets, we have air bubbles. Since the gas does not perceptibly contribute to weight, foams feel light when we eat them.

Gas can have another benefit: it can contain volatile aroma molecules. Chefs imaginatively use this property as a way to incorporate volatile aromas to create innovative sensory eating experiences. Jordi Roca had one of his first creative breakthroughs when he discovered ice cream's extraordinary ability to absorb flavors and aromas. He utilized the fact that ice cream is a foam, and that the many, many air bubbles inside it together create a huge surface area through which flavor molecules can be absorbed. This gave rise to the idea of a cigar smoke–flavored ice cream and introduced a whole new concept to the world of pastry chefs.

Note that in the recipe, to incorporate the cigar smoke aroma, Jordi attaches and lights a cigar at the end of a manual air pump, and then pumps the air into the churning ice cream base. As the ice cream solidifies, the air containing the aroma molecules is trapped within the air bubbles, only to be released when the diner takes a bite. To complete the dish, appropriately named "A Trip to Havana," the ice cream is served inside a rolled up sheet of chocolate, making it look almost indistinguishable from a real cigar, complete with ashes—a spice and sugar mix—and an ashtray.

Cigar Ice Cream

Ingredients
Coal candy (aka "sweet coal," a confection
   made from sugar and egg white that looks
   like lumps of black coal)
Tempered Chocolate Cylinder (recipe follows)
Lime Gel (recipe follows)
Mint Slush (recipe follows)

Rum Candies (recipe follows)
Cigar Base (recipe follows)
Mint flowers

Directions

1. With a Microplane, grate the sugar coal and sprinkle a small amount in a cigar ashtray.

2. Dip the tip of the chocolate cylinder that hasn't been covered by dark chocolate into the coal to simulate a cigar with ash at the end.

3. Place the cigar on the side of the ashtray.

4. Fill a shot glass three-quarters of the way with lime gel, then arrange the mint slush and rum candies on top. Place the mint flowers on the surface.

Rum Candies

Ingredients

60 g sugar
20 g water
12 g rum
Cornstarch

Directions

1. Prepare a simple syrup by combining the sugar and water and bring to 228°F (109°C). Remove from the heat and let cool slightly. Carefully, pour the rum into the mixture. The ideal way to do this is by transferring the syrup slowly back and forth from one container into the other, letting it slide until both liquids and densities combine.

2. Preheat the oven to 176°F (80°C).

3. Cover a rimmed baking sheet with cornstarch and put it in the oven until dry (at least 3 hours). Turn the oven temperature down to 104°F (40°C). Press down on and smoothen the cornstarch, then press small holes into it; these will serve as molds for the rum mixture. Add the rum mixture to the small holes, sprinkle the surface slightly with some more dry cornstarch, and bake for 24 hours.

4. Remove the candies from the cornstarch, brush clean, and reserve.

Lime Gel

Ingredients
75 g water
20 g sugar
1.2 g agar-agar
20 g lime juice

Directions
1. Bring the water and sugar to a boil in a small saucepan until the sugar dissolves.

2. Let the syrup cool, then add the agar-agar and again bring it to a boil.

3. Remove from the heat. Stir in the lime juice, then transfer the mixture to a mold to cool and solidify.

4. When curdled, blend the mixture to a silky gel.

Mint Slush

Ingredients
25 g water
25 g dextrose
12.5 g inverted sugar syrup
½ gelatin sheet, bloomed (soaked in cold water for 5 to 10 minutes and drained)
100 g Mint Water (recipe follows)

Directions
1. In a saucepan, combine the water, Mint Water, dextrose, and inverted sugar syrup and bring to a boil.

2. Remove from the heat and add the gelatin.

3. Blast-chill or place in the freezer for 30 minutes.

4. Mix the cold syrup with the mint water and freeze.

Mint Water

Ingredients

50 g cold blanching water

75 g mint leaves

50 g room-temperature water

Directions

1. Bring a pot of water to a boil. Blanch the mint leaves for about 20 seconds, then transfer to the cold blanching water.

2. Add the room-temperature water and blend to a fine, smooth liquid. Run through a fine chinois and press to obtain as much mint water as possible.

Tempered Chocolate Cylinder

Ingredients

200 g dark couverture chocolate, tempered (see chapter 2, sidebar 10)

Cigar Base (recipe follows)

Directions

1. Cut 5 cm by 10 cm rectangles of grease proof paper.

2. With a palette knife, spread a very fine layer of the tempered chocolate on each rectangle, leaving 1 cm free on one end.

3. Roll the paper into cylinders so that the clean end of the paper sticks out like a flap that can be pulled to remove when serving. Leave the chocolate to dry.

4. Fill the chocolate cylinders with cigar ice cream base with a piping bag.

5. Freeze, then dip one end of each "cigar" in the same couverture chocolate.

Cigar Ice Cream Base

Ingredients
750 g heavy cream
150 g dextrose
4 gelatin sheets, bloomed (soaked in cold water for 5 to 10 minutes and drained)
1 Partagás Serie D No. 4 Cuban cigar

Directions
1. Mix the cream and dextrose in a saucepan and bring to a boil.

2. Remove from the heat and add the gelatin, stirring well.

3. Refrigerate for 6 hours.

4. With a mixer, whip the cold cream mixture while smoking the cigar next to the bowl.*

5. When whipped, remove the ice cream and chill.

The Rocas have a special air pump for this purpose, but smoking the cigar the tradi-tional way also works. ❀

Foams consist of gas bubbles dispersed in a second phase. The continuous phase can be a liquid or, when cooked or gelled, can become a solid. Just as the drops of an emulsion must be stabilized against coalescence, so too must the bubbles in a foam. As in emulsions, this stabilization is achieved through the use of surfactants, emulsifiers, or solid particles. And just like oil or water drops, deforming the bubbles is like deforming an elastic solid. As a result, foams with high volume fractions of air, such as the head of a glass of beer, are solids. While emulsions create solids out of two dense liquids, foams are remarkable in that they mix a liquid and air to form a solid.

One major feature of foams arises from the fact that the liquid continuous phase is always much denser than the gas phase in the bubbles. As a result, liquid foams continually drain because the fluid sinks to the bottom and the bubbles rise to the top. This increases the volume of the bubbles at the top, making them more solid. The head of a beer is a stiff, dry foam, while the liquid beneath is a fluid that contains some small bubbles. Upon pouring, the foam forms at the top within seconds because the liquid beer has a rather low viscosity. If we continue letting the foam sit, more fluid will drain out from between the bubbles, making them drier and more brittle. Some types of beer add specific molecules that slow the drainage of the last bits of liquid, making the head on these beers last much longer, and giving them a very different texture. Textural changes due to drainage explain why the culinary foams presented in haute cuisine restaurants must often be eaten quickly.

WHIPPED CREAM

Whipped cream is traditionally a particle-stabilized foam. It is made by whipping cream until it forms a solid foam. The whipping causes fat globules to break apart and aggregate at the surface of the air bubbles, where they help stabilize the bubbles. A common type of whipped cream is sold in pressurized cans that contain nitrous oxide (N_2O), which exists as a liquid inside the pressurized canister. When you shake the can, the liquid nitrous oxide forms an emulsion with the liquid cream, and as soon as the cream is dispensed, the nitrous oxide, now at normal air pressure and temperature, turns into a gas. The small liquid drops in the emulsion rapidly expand, and you now have cream with bubbles, or whipped cream. Compared to the bubbles you get from traditional whisking techniques, the bubbles in these kinds of foams are minuscule. They also do not rely as much on emulsifiers, so it is possible to turn a wider range of ingredients into airy, whipped foods with this mechanism. But there are also disadvantages. For example, hand-whipped cream has fat globules on the surface of the bubbles, which significantly affect the taste; it will always taste milkier and richer than canister cream. In addition, the fat globules are much better stabilizers of the foam; thus a hand-whipped cream remains stable and tasty for much longer than whipped cream formed with nitrous oxide. As you can imagine, both methods have their uses in the kitchen.

ANGEL FOOD CAKE

Foams made with eggs have a useful property: we can turn them into a gel simply by heating them until the protein network is irreversibly solidified. When the continuous phase is a liquid, heating the foam causes the gas in the bubbles to expand; this is why breads and cakes rise in a heated oven. However, simultaneously, heat causes the continuous phase of a foam to become a solid. In the example of bread, wheat proteins are denatured by heat, and they set into a solid network around the air pockets in the dough. Pastry chef Joanne Chang makes an angel food cake in which the egg proteins solidify during cooking. This solidification causes the foam to feel firm. Many bread recipes call for an initially high baking temperature, which is then lowered partway through baking. This allows the air bubbles to expand as much as possible before the proteins set. The solid properties of cooked foams like bread or cake have much more to do with the solidified continuous phase than with the packing of gas bubbles, although recipes for solid foams often require the initial formation of a liquid foam. Since cooked foams have a sponge-like texture, they are excellent for soaking up sauces or spreads. In Italy this practice even has a name: *fare la scarpetta.*

ICE CREAM: A FOAM, AN EMULSION, AND A COLLOIDAL SUSPENSION

Like cookie dough, ice cream is both an emulsion and a foam. The ice cream base consists of water, milk fat and proteins, and sugars. These ingredients work together to create a creamy, smooth dessert once frozen together. There are many dispersed phases. One of them is made up of milk fat globules suspended in water. Another is the air, which is incorporated in the form of bubbles during the churning process. The milk proteins can help stabilize the emulsion as well as the air bubbles. There is one other component that is important for ice cream: the ice crystals. Scientists call this type of dispersion, where small solid particles are dispersed in a continuous phase, a colloidal suspension. The ice crystals are also created by the churning process, but their size is controlled by the sugar in the recipe.

The sugar lowers the freezing point of water, which both keeps the crystals small and prevents the water from fully freezing into a hard solid. In fact, the ice crystal size is inversely related to the perceived creaminess of the dessert, so a good ice cream must have small ice crystals. In the ice cream you made in chapter 2, the size of the ice crystals was determined by how rapidly you churned the ice cream, which happened by kneading it in a plastic bag. Professional ice cream makers produce creamier ice cream by having more sophisticated churning mechanisms.

As a final treat, we leave you with a recipe that controls the sizes of the ice crystals in a different way: liquid nitrogen ice cream. This way of making ice cream has been popular in recent years, probably at least in part because the cold smoke looks cool. It is also remarkably creamy, and this is because of the very small ice crystals. Liquid nitrogen is only a liquid at temperatures below −196°C (−321°F); it is colder than anything else we ever work with in a kitchen. Because it is so cold, it is dangerous to work with if you are not careful. However, the advantage of using it to cool down an ice cream base is that there is almost no time for the water to form crystals. After the base has partially frozen, churning is still necessary to incorporate air for a lighter texture. Together, these two steps create ice cream with a wonderfully smooth texture.

SIDEBAR 6: LIQUID NITROGEN ICE CREAM

Liquid Nitrogen Ice Cream

Ingredients
90 mL heavy cream
100 mL milk
20 g sugar
Pinch salt
Flavoring (optional)
Liquid nitrogen*

Directions

1. In a bowl, mix together the cream, milk, sugar, salt, and any flavorings you choose.

2. Pour a small amount of liquid nitrogen directly into the bowl.

3. After a few seconds, start mixing with a wooden spoon or spatula to lighten the ice cream. If you are using a stand mixer, start on the lowest speed and occasionally scrape down the sides. The goal is to freeze the ice cream evenly while incorporating air.

4. When the smoke is gone, all the liquid nitrogen has evaporated.

5. Pour in more liquid nitrogen as necessary, mixing until the desired texture is reached.

A note on safety: Liquid nitrogen is very cold and can cause severe damage to your eyes, nerves, and cells, so proper safety protocols must be followed. When working with liquid nitrogen, always wear a face shield and cryogenic gloves, and make sure no skin is exposed. In addition, wear nonabsorbent clothing. Although a small splash may not cause any harm (thanks to the Leidenfrost effect, which occurs because the liquid nitrogen rapidly vaporizes, forming a thin layer of air between it and your skin, thereby protecting you from a few drops), absorbent clothing will hold on to the liquid nitrogen, giving it more time to contact and damage your cells. Finally, only try this recipe in a well-ventilated area. The sudden introduction of so much nitrogen gas into the air can decrease the oxygen concentration nearby to below safe levels. ⚛

CHAPTER 7

Microbes

Microbes are essentially invisibly small cooks that change foods very much for the better, their physical properties included.

—Harold McGee

Cooking with Life

Think about what you've eaten today and ask yourself if you've ingested anything that was "cooked with microbes." If you did *not* say yes, consider the following: Did you have coffee first thing in the morning? Yogurt for breakfast? A sandwich with bread, cheese, or a pickle at lunch? Or how about beer or wine after a long day at work? Are you perhaps nibbling on a piece of chocolate right now? Or did you have soy sauce, kimchi, vinegar, or tempeh? If you had any of these foods, you should have answered yes, because all of them have been cooked with microbes.

Fermented foods—this is how we usually refer to foods that have been produced with the help of microbes—are much more ubiquitous than most people realize. Many of us eat them every day, often without knowing how they are made or that they may have been produced with yeast or even bacteria, which we often shun as having to do with disease and bad hygiene. Before you get too grossed out by this, let us first assure you that the microbes of food fermentations are very different from disease-causing microbes. On the contrary, they can be quite helpful in counteracting them. Moreover, this way of cooking is not some unproven modern invention. It goes far back in our culinary history. The earliest evidence dates back some 9,000 years ago, in the form of chemical remains of mead in ceramic vessels found in China and bones of fermented fish in Sweden. But fermentation practices likely go back much longer than this. Fermentation's history is also widely spread across cultures, including foods such as leavened bread and cheese in Egypt, fermented soybeans in China, and many others. Cultures around the world have engaged in food fermentations of various kinds for a long time, and still do.

The reason for the early prevalence of fermented foods in human history is the impressive preservative properties that can be achieved with the right microbes. In the time before refrigerators and freezers, fermentation was a critical method to ensure a healthy food supply. But cooking with microbes also bestows many other wonderful properties on foods: new flavors and textures. In this way, cooking with microbes is not so different from cooking with heat or any of the other processes we have seen so far in this book. It is a marvelous method for transforming food.

But let's stop for a minute and think about this. Bacteria and yeast are *alive*. This stands in stark contrast to all the other cooking techniques we have seen so far. The previous processes, in spite of all of their glorious complexities, were all, in one way or another, using nonliving things. In fact, several of them were specifically designed to kill any life that may have been in the food. Here we are doing something entirely different—we are, so to speak, cooking with life.

Let's explain what we mean by this. The cooking processes so far in this book have tampered with proteins in various ways, whether by applying heat, adding salt, altering the pH, or even whisking. We said that the proteins "cooked," which, on a microscopic scale, meant that they denatured and became nonfunctional. This view of cooking has an interesting implication: All life depends on the function of proteins. In fact, all of the chemical reactions that are necessary to keep biological organisms alive are executed by the vast number of proteins in the cells of those organisms. As a consequence, most cooking methods not only destroy the proteins, they also kill the organism. For example, if you put yourself in a very hot bath—say, 65°C (149°F), like the ones we used to cook sous vide steak—eventually the proteins in your body will denature, and you will die. The same is true when cooking an egg. By cooking an egg, the life that was in the egg is destroyed.

So, crassly speaking, we have just spent a considerable number of pages trying to destroy proteins, and thus indirectly life itself. But in this chapter, we will do the exact opposite. We do not want to damage the proteins. Instead, we want to keep them highly functional. The reason is that in this chapter, the proteins, and indeed the organisms that hold them, are doing the cooking for us.

Bread, Mead, and Sauerkraut

You may have eaten fermented foods, but have you ever made a fermented food yourself? If not, we highly recommend it. In fact, how about you try one (or all) of the recipes in the sidebar right now. All are relatively straightforward. Then, as the microbes do their job transforming the food, come back to this chapter and read about the science of the soon-to-be-bubbling jars on your kitchen counter.

SIDEBAR 1: BREAD

Bread

Ingredients

2 cups (480 g) lukewarm water

1 teaspoon sugar

1 (¼-ounce/7 g) package active-dry yeast

4 cups (480 g) all-purpose flour

1 teaspoon salt

Directions

1. Pour the lukewarm water into a large bowl and stir in the sugar and yeast. Within 5 to 10 minutes, there should be bubbles and signs of activity from the yeast.

2. Add the flour and salt and mix well, kneading by hand or with a stand mixer fitted with a dough hook for 2 to 3 minutes. The dough should be wet, loose, and shaggy, and no clumps of flour should remain.

3. Cover the dough with a damp towel and let rise in a warm, draft-free location for 1 hour, or until doubled in size.

4. Preheat the oven to 425°F (218°C). Lightly flour a rimmed baking sheet

5. Punch down the dough and divide it in half. Shape into two loaves (rounds, batards, or whatever shape you like). Place the loaves on the floured baking sheet, cover, and let rest for 30 minutes.

6. If desired, score the top of the loaves with a sharp knife. Bake for 15 minutes. Turn the oven down to 375°F (191°C) and continue baking for 10 to 15 minutes, until the crust is golden brown.

7. Cool on a wire rack for at least 20 minutes before slicing. ✸

SIDEBAR 2: MEAD

Honey

Water

Yeast

Mead

Ingredients

12 cups water

3 cups raw honey

1 (5 g) packet champagne or wine yeast (optional but recommended)

Directions

1. Combine the water and honey in a crock and stir until the honey is dissolved. Add the yeast. (Note that you can leave out the yeast altogether—naturally occurring microbes will still ferment the mead, it will just take a little longer.)

2. Cover the crock with a towel and place in a warm location for 3 to 4 days, stirring several times a day.

3. When bubbly, transfer the mixture to a clean jar.

4. Cork with an airlock, or *very* loosely with a lid so that pressure cannot build up.

5. Ferment for 2 to 4 weeks, taste-testing regularly to decide when you consider it done. The finished mead should be fizzy and still sweet. If you let it ferment for too long, the microbes will consume all the sugar and the mead will no longer be sweet. ⚛

SIDEBAR 3: SAUERKRAUT

Sauerkraut

Ingredients

½ head green cabbage, finely chopped,
  plus several loose leaves

Salt

Cabbage
2 pounds

Salt
3 teaspoons (2%)

Directions

1. Weigh the cabbage. Calculate 2% of the weight for the salt.

2. Place the cabbage in a large bowl and sprinkle the salt on top. Toss to distribute the salt evenly throughout the chopped cabbage.

3. Massage the cabbage to break up the structure and help release the water. The cabbage should "sweat" and release a substantial amount of brine.

4. Transfer the cabbage and liquid to a mason jar. Add a little cabbage at a time and pack it tightly with your fist or a kitchen utensil such as a pestle, making sure to push all the air out. The brine should cover all of the cabbage.

5. Fold several cabbage leaves and place them on top of the chopped cabbage. They will serve as a lid, and you can peel them off one at a time if mold develops. Make sure the cabbage is submerged under the brine. You can place a weight, such as a small dish or a clean rock, on top to push it down.

6. Loosely cover the jar with a lid and leave to ferment at room temperature for 1 to 4 weeks. Be sure to "burp" the sauerkraut daily—open the lid to let air out—to prevent a cabbage juice explosion. Taste the sauerkraut every few days to determine when you consider it done.

7. Store, covered, in the fridge. ✻

Let's start by taking a look at the bread recipe because it's a process many of us are somewhat familiar with. The only ingredients are water, flour, perhaps a little sugar and salt, and, of course, yeast. The instructions are similarly straightforward: you mix all the ingredients together, place the dough in a warm spot, and then wait for it to rise. If you have baked bread before this is not new to you, but let's admire this process and think about what it may tell us about what's going on.

The dough starts out dense and small, but over time it grows, becoming airy and large. Watching it, you can almost tell it is alive! How did this happen? One of the things we have seen in this book is that by cooking and experimenting in our kitchens, and simply asking questions about what we observe, we can learn a great many things. This recipe is no different. There are all kinds of questions that arise even from these simple ingredients. Why do you add sugar—is it because you want your bread to be sweet? Why do you put the dough to rise in a warm spot? Why do you have to wait for so long for the rising to happen? In this chapter, the answers to our questions are very often that the microbes did something. And indeed, this is also the case in the bread.

Here we added yeast, so it seems reasonable to assume that the yeast is responsible for the transformation. The yeast produces a gas, which spreads and fills throughout the dough. The structure of the dough itself is strong enough to trap the gas, preventing it from dissipating in the room; this makes it seem as if the dough is getting bigger and bigger. Let's leave the dough to rise for a while and take a look at the second recipe.

Mead is even simpler than bread. You just mix water, honey, and yeast in a jar. Simplicity is a hallmark of many food fermentations, and it distinguishes them from many other types of cooking, which can be quite complex. For fermentations, often all you need are a few ingredients and, with some time—perhaps the most important ingredient of all—the microbes will do the cooking for you. We've seen how the control of time is a critical ingredient in all kinds of cooking processes, from heat diffusion in steak to lemon juice cooking ceviche. In this case, we need a long period of time: days, if not weeks and months. The mead recipe specifically calls for 2 to 4 weeks.

What's happening here? If you have already started this recipe, you know that the liquid becomes a bit fizzy—it bubbles. Just as with bread, a gas is produced. If you've ever tasted mead, you know that the final product is alcoholic, and somewhat sweet, but not overly sweet. From these facts alone—the disappearance of sugar and the appearance of alcohol and gas—it seems reasonable to assume that the sugar has somehow been converted to alcohol and gas. And if you guessed that the guilty offender for this transformation is the yeast, then you were entirely correct. Yeast loves sugar. Anywhere with abundant sugar there will be yeast. And both the mead and bread recipes contain plenty of sugar, in the form of simple sugars in the honey and carbohydrates in the flour.

The bread and mead recipes are basically the same process. The only real difference is that for bread the gas is trapped in the matrix of the dough, whereas for mead the gas is allowed to escape. Similarly, for mead we let the process go longer so it produces more alcohol, whereas for bread we produce only a small amount of alcohol, which evaporates during baking. But both recipes feature the same type of microbe. In fact, the yeasts are so similar that they even have the same name, *Saccharomyces cerevisiae*. If this is a surprise to you, compare the ingredient lists on the baking yeast in your kitchen and the little pouch of champagne yeast you may have bought from a brewer's store to do the recipe in the sidebar—both will have the same Latin name. It is only when you look closely, at the subspecies level, that some differences emerge: yeasts used for mead and wine tend to produce more alcohol and aromatic flavors than bread yeast. This is an important difference if you are a winemaker, for sure, but not significant enough for the microbes to be considered different species scientifically. In fact, you can make acceptable mead with baker's yeast and, conversely, even bake a delicious loaf of bread with champagne yeast—just try it and you'll see.

But enough of bread and mead for now, let's check in on our third recipe: sauerkraut. You make sauerkraut by chopping cabbage, then massaging and squeezing it with salt before firmly pressing it into a jar. The microbes themselves are conspicuously absent in this recipe. You're not adding a package of yeast as you did in the bread and mead recipes. But the apparent effortlessness of the recipes belies both the underlying scientific complexity as well as the fact that microbes have a dramatic effect on the final food. If you started this recipe and already tasted it, you know that the relatively flavorless cabbage has turned deliciously tangy. The sour flavor has been produced by lactic acid bacteria. These microbes also love sugars, which they get from the carbohydrates in the cabbage, but they make lactic acid instead of alcohol and gas as the yeast did.

Oxygen

Take a deep breath. Breathe in the air and fill your lungs. Then slowly breathe out. Do this a couple times and think about this process, which you are doing every moment of every day of your life. What are you breathing in, what's coming out, and what's happening in between? In simplified terms, oxygen comes in and carbon dioxide goes out, and in between your body manages the not insignificant task of breaking down the food you had for lunch and extracting energy from it. You need the energy to live. It fuels all the cellular processes in your body. We call this process respiration, and the all-important molecule required for it is oxygen.

Now, take another deep breath, but instead of breathing out, hold your breath. Keep holding it, and holding it, for as long as you can.

How's it going?

Not so well?

At some point you'll give up and take a deep breath in desperation. The air's oxygen is so important that you cannot do without it.

Now, if you were a microbe, the story would be entirely different. In fact, the reason we're asking you to do this exercise, inspired by our friend and colleague Roberto Kolter of Harvard Medical School, is that it perfectly illustrates what goes on in your mead and sauerkraut. If you were a microbe, you would, when told to hold your breath, simply switch to another mode of metabolism that doesn't require oxygen. You may even prefer this way of "breathing"! At least this is the case for the microbes in the mead and the sauerkraut. Recall how we tightly packed the cabbage in the jar? This was to get rid of all the air and make the environment free of oxygen. We call this oxygen-free mode of metabolism "fermentation." It basically comes in two forms: alcohol production by yeast and lactic acid generation from bacteria. These two types of food fermentation are so common that the term "fermentation" has come to refer to any process where food is produced with the help of microbes, not just the cellular process it was originally referring to.

The reason these two food fermentations are so prevalent is not because of their oxygen-free metabolism, but rather because of the waste molecules they produce. Both acidity and alcohol are toxic to most microbes. The super power of lactic acid bacteria is that they are able to tolerate small amounts of acidity just slightly better than other microbes. So even if there is less than 1% of them on the fresh cabbage, the acid they produce will make the environment just slightly more hospitable for them, while simultaneously making it less ideal for other microbes.

This makes the lactic acid bacteria grow better, which makes the environment even more acidic, and this goes on and on. Slowly, the lactic acid bacteria have created an environment where they can dominate and hog the food source for themselves. Quite the super power indeed. Even better, for us humans the food is now "preserved" because any spoilage microbes cannot grow. The alcohol in mead works in much the same way.

Microbes

What is a microbe? We can guess at, or be told of, their existence. But to see is to believe, and we haven't yet seen any microbes, have we? Only their effect. Let's see if we can spot them. Imagine diving into the sauerkraut on your kitchen counter, while simultaneously zooming in. You'd watch as the cabbage pieces get larger and larger, the individual cabbage cells emerge, occasionally air bubbles would rise around you, and finally, after having zoomed in much further, there they are—the microbes! Each microbe consists of only one cell. In sauerkraut they are slightly smaller than what you'd see if you dove into the mead or bread; bacteria are supposed to be smaller than yeast, so this makes sense. But both are tiny. If one of your hairs had accidentally fallen in when mixing and squeezing the

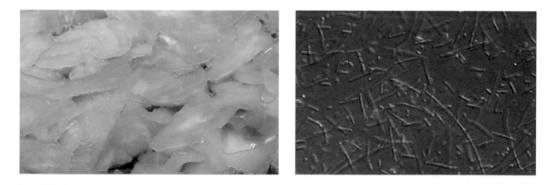

FIGURE 1 **(A)** Close-up of sauerkraut after a few days of fermentation. Carbon dioxide bubbles are collecting in pockets, the texture is softening, and the color is changing from bright green to yellowish green.

(B) A drop of liquid from the sauerkraut as seen under a microscope at 1,500 times magnification hints at the diversity of microbes, from small and round to thin and elongated. Microbes that appear to be attached to each other are undergoing division.

cabbage, you would now, with your magnified view, see that you'd be able to fit up to 10 or 100 microbes across the breadth of that hair.

As important as food fermentation has been for most of human history we've had no idea what these creatures looked like, nor that they even existed. It wasn't until the late 1600s that a Dutch scientist by the name Antonie van Leeuwenhoek got to see the kind of zoomed-in picture of microbes that we just saw in our sauerkraut. As a textile merchant, Leeuwenhoek had perfected the art of grinding lenses to help him see the quality of fabrics in detail. One day he decided to put a drop of Delft canal water under the lens. What he saw must have been astounding—multitudes of tiny creatures moving around! A new world, just on a smaller scale. And not so different from the world in our mead and sauerkraut.

Growth Conditions and Exponential Growth

As you imagine yourself swimming around in the sauerkraut, there is a second thing you'll notice after marveling at how tiny the microbes are: there are so many of them! Where did they all come from? Well, for the mead and bread recipes, the answer is simply because we added them as one of the ingredients. But we didn't add any microbes to the cabbage and, according to the mead recipe, the yeast package wasn't necessary there either.

Microbes are everywhere. Since we don't see them, we don't notice them. This is true not just for the microbes that we can cook with, but for all kinds of microbes, and there is a whopping number of different ones. Microbes get a bad rap because of a few dangerous examples, but in general they are harmless, often even beneficial, and we live with them on a daily basis in the air, in our guts, and on the materials and surfaces around us. They were in van Leeuwenhoek's canal water. And they are on cabbage leaves, in honey, and in flour.

In the bread and mead recipes, these natural microbes don't stand a chance because we're adding millions of yeast cells (yes, this is how many there are in a teaspoon). This gives them a big advantage, and they quickly outcompete anything else that might grow. But we could have relied on the natural microbes. If given the right environment, these microbes will grow and take over. Even a tiny advantage will help. This is what happened in our sauerkraut. And it's what could have happened in the mead if we had not added any yeast. The key is to create the right growth conditions.

GROWTH CONDITIONS

Let's think back to the bread dough, which you were told to put in a warm spot to rise. That's because for the yeast, as with all organisms, there is a set of environmental conditions where it can grow and thrive best. We've already encountered some of the ones that are important for microbes: temperature, pH, salt concentration, and the presence of oxygen. Looking at almost any fermentation recipe will reveal how these parameters are being optimized. Each type of microbe has some optimal amount for each condition, as well as minimum and maximum amounts outside which the microbe cannot live.

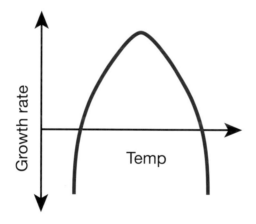

FIGURE 2 The graph shows the growth curve for an imaginary species of microbe at different temperatures. There is an optimal temperature at which the microbes grow at maximum rate. Above and below this temperature, growth is slower. At some point, it is either too hot or too cold for the microbes to grow at all and they start to die off—this is indicated on the diagram as the temperature at which the growth line crosses the horizontal axis and becomes negative. Microbes have growth curves for all of their growth conditions—you can imagine similar curves for pH, oxygen concentration, and salt concentration. For each condition there is some preferred optimal amount, and a general zone on either side that is tolerable for growth. In combination, the growth curves tell us the microbes' preferred environment.

SCIENCE AND COOKING

Figure 2 shows growth rate as a function of temperature for an imaginary microbe. You can see that there is an optimal temperature, and then growth decreases in both directions until it is not happening at all. Outside of these bounds the growth rate is negative and the microbes have not only stopped growing; they are dying.

We make use of these kinds of curves all the time when cooking, whether directly or indirectly. At the low temperature extreme are refrigerators and freezers, where it's too cold for microbes to grow. Similarly, pasteurization and sterilization occur at the high temperature extreme, where the microbes die. Food safety is all about staying in the right regions of these growth curves. The same is true for fermentation recipes. All microbes—and, in fact, all organisms—have growth curves like the one in the figure, not just for temperature but for many other important conditions. The yeast in the mead have one for oxygen, sugar, and temperature, among others. The lactic acid bacteria depend on salt, pH, oxygen, and temperature. When fermenting, we manipulate one or a combination of these conditions to coax particular microbes to grow more than others. By doing so we can make delicious food. The sidebar shows how this plays out in a recipe for sourdough bread, an age-old food fermentation that is the precursor to all bread baking of today.

SIDEBAR 4: SOURDOUGH STARTER AND MICROBIAL COMMUNITIES: GROWTH CONDITIONS AND COMPETITION

We've gotten to know the two main players of food fermentations: yeast and bacteria. Sourdough is a food where they get to mix. It is also a fantastic example of what competition within a microbial community can look like.

Sourdough Starter

Ingredients

500 mL flour (any kind), plus more as needed

500 mL water, plus more as needed*

6 organic grapes or berries, unwashed (optional)

1 pinch yeast (optional)

Tap water is fine, but if it is heavily chlorinated, leave it out overnight. Water from cooking pasta or potatoes can also be used.

Directions

1. In a bowl, mix the flour, water, and yeast. Add the grapes or berries, if using—they will help introduce wild yeasts to the starter—and stir vigorously.

2. Cover the jar with cheesecloth and set aside in a warm location. Stir once or twice daily until the batter is bubbly, 1 to 2 days. Strain out the fruit.

3. "Feed" the mixture daily with 1 to 2 tablespoons additional flour and continue stirring daily. Add small amounts of water as needed to maintain a liquid batter.

4. After 3 to 4 days, when the starter is thick and bubbly, it is ready to use. Remove some starter as needed for bread baking. Save some starter and continue feeding it equal parts flour and water. With continued feeding and care it will keep indefinitely! ⚛

Sourdough starter can be made from scratch by mixing flour and water. Over time, the yeast and bacteria on the flour and in the air grow and create a unique environment. The result is an interesting and delicate interplay between the two. Knowing what you know about these two microbes you can appreciate why. Yeast eat sugar and produce gas; this is what will eventually make the dough rise. The bacteria also eat sugar, but they produce acid. Yeast do not like acid. Additionally, the bacteria divide much faster than the yeast, so the bacteria will easily take over, making the environment too acidic and outcompeting the yeast. The result is bread that doesn't rise. This is not the outcome you want. If, on the other hand, the yeast take over, you get none of the desired sourdough tang. Also bad. The optimal ratio for rise *and* acidity is 1 yeast per every 100 or 1000 bacteria. The trick to getting there is to work with the growth conditions: bacteria like slightly warmer temperatures than yeast. Keep your sourdough starter on the cooler side, and the yeast will get the small advantage they need. In other words, by optimizing temperature, one of the yeast's growth conditions, it can tolerate a less ideal pH, another of its growth conditions. With this small tweak, you have engineered an environment that leads to delicious food. The balance is delicate, though, and even small changes to the environment can throw it off. The picture is further complicated by the fact that there are many different types of yeast and bacteria, all with slightly different preferences. They contribute to the unique flavor profile of sourdough breads from around the world. We are just starting to learn more about their enormous microbial diversity. Guess where *Lactobacillus sanfranciscensis* is from? (Hint: reread the second word of the name.)

EXPONENTIAL GROWTH

We have now created the right growth conditions and we're waiting for our mead, sauerkraut, and sourdough bread to "cook." But what are we waiting for, really? The answer to this question holds the key to all food fermentations. As you imagine yourself swimming around among the microbes in your sauerkraut, you'll notice that they are multiplying. The bacteria grow longer and longer until they split in half and become two. The yeast don't split in half but instead grow little buds that detach and become new yeast cells. Just imagine if you could do that.

Yeast and bacteria are terrific at multiplying themselves, and this is partly what makes them such great cooks. How else would these tiny creatures make all that alcohol? If the conditions are right, the yeast in your mead will make a new yeast cell approximately every 2 hours. Many bacteria are even faster, sometimes dividing as often as every 20 minutes. By comparison, humans are excruciatingly slow. It takes an average of 20 to 30 years to create another human being.

Generation number	Number of bacteria
0	1
1	2
2	4
3	8
4	16
5	32
6	64
7	128
8	256
9	512
10	1,024
11	2,048
12	4,096
13	8,192
14	16,384
15	32,768
16	65,536
17	131,072
18	262,144
19	542,288
20	1,048,526
21	2,097,152
22	4,194,304

FIGURE 3 Microbes grow exponentially. If one microbe divides into two new microbes, and each of these two microbes also divide into two new microbes, you will get a total of four microbes after two generations. If this continues, the total number of microbes will quickly become very, very large. This is the power of exponential growth; even small numbers of microbes can grow into huge populations as long as they are given unlimited food and optimal growth conditions.

If you're not impressed by this fast generational turnover, take a look at the numbers in the figure. Let's say you start with one microbe and it divides into two, then each of the two microbes divide into two, and so on and so forth. You get huge numbers very quickly. Mathematicians call this process exponential growth. It can be summarized in a formula, but you don't need an equation to be impressed by the numbers that result from it. You can see them in the figure. Let's say there was one "bad" microbe in your lunch today. It is now evening, so this lone cell and its descendants have been dividing for the last, say, 7 hours. With each generation taking 20 minutes, this means there have been 22 divisions total, which is a whopping 4 million microbes. Even if each microbe is tiny, together these 4 million microbes weigh several grams. Not good news for a good night's sleep.

Let's say we wait for a little longer, say, 44 hours, almost two whole days. Now the microbes weigh as much as our planet! At least if we believe the exponential equation. Should we?

Lucky for us, the situation isn't as bad as this. *If* the microbes had enough food, and could continue to divide at the same rate, then yes, this would happen. But there is never enough food. Only in the very beginning of, say, a mead recipe, do we see growth like this. Very quickly the growth levels off, and when the sugar is gone, the yeast die. So there is no risk we'll be overtaken by the bad microbes in your lunch anytime soon, at least not this way, or this easily. Nevertheless, the possibility for exponential growth is the reason one microbial species can so quickly take over and dominate. Without it, none of the fermented foods we know and love would be possible.

PREDICTING THE MAGIC

At the surface, food fermentations can seem both complicated and a little mysterious: Tiny, invisible organisms transform food molecules into flavor molecules. They do this with intricate molecules called enzymes in a system of convoluted chemical reactions. Is there anything systematic about this process? Can we predict what will happen without understanding all of the steps? Yes, we can. The trick is to focus only on what goes in and what comes out, and ignore the steps in between. Let's see how this works with one of our favorite fermented foods: the transformation of grape juice to wine.

SIDEBAR 5: GRAPE JUICE TO WINE TO VINEGAR

Let's say you have enough grapes to make one bottle of wine (750 mL or 750 g). The sugar content of your grape juice is about 20% by weight, although this will vary slightly depending on ripeness and other factors. You know that the yeast eats the sugar and converts it to ethanol and carbon dioxide. We can write this as a reaction where each sugar molecule—in this case we assume that all of the sugar is glucose—makes two ethanol molecules and two carbon dioxide molecules. If the yeast eat *all* of the glucose molecules (that is, 20% of 750g, which is 150 g), we would get a little more than 100 mL ethanol. This corresponds to 13% ethanol by volume (100 mL/750 mL). Go take a look at the alcohol content of any wine bottle in your house and tell us what it is—we

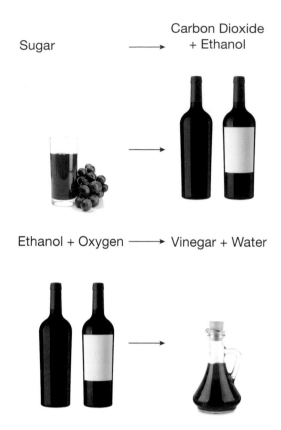

Sugar $\longrightarrow$ Carbon Dioxide + Ethanol

Ethanol + Oxygen $\longrightarrow$ Vinegar + Water

bet it's within or close to this range! The reason, of course, is that the alcohol content of wine is limited by the sugar content in the grapes. Since the sugar content varies by only so much, the alcohol content in different wines doesn't vary much either. We can thus predict the final alcohol content simply by looking at what the yeast ate.

In this example, the yeast more or less ate the sugar until it ran out—you can tell because most wines are not sweet at all. But what if you wanted to make a sweeter wine? Or one with higher alcohol? Well, you would just add more sugar. Rather than simply pouring in table sugar, wine makers have creative ways of doing this. One way is to let the grapes sit and dry on the vine until they are almost raisins, which concentrates the sugars. Another is to not pick the grapes until after the first frost. This also increases the sugar content and is how ice wines are made. Yet another way is to inoculate the grapes with a mold, a common practice for making Tokajis in Hungary or Sauternes in France.

But if instead of increasing sugar, you wanted a higher *alcohol* content in your wine, you would run into trouble. Ethanol in high quantities is toxic, and even the hardiest yeast strains cannot tolerate alcohol concentrations much above 20%. Thus any wine with a higher alcohol concentration than this would either have to be distilled, like whiskey, rum, and vodka, or be fortified with additional alcohol as is the case with port, sherry, and vermouth.

Let's say you've produced your delicious wine. It is bottled and ready to be aged for many years, over time becoming even more delicious. The alcohol will preserve it from any spoiling microbes, so now you can relax and just wait, correct? Yes, mostly this works. But you can't be too sure. In the world of microbes, there is almost always some bug that can survive on food sources where nothing else can grow. This is also the case for wine. Have you ever excitedly opened a very old bottle of wine only to realize that it had turned sour? If so, you've experienced the results of acetobacter and gluconobacter, two bacteria that eat ethanol the way yeast eat sugar. They extract energy from it and, in the process, produce vinegar. Unlike yeast, they require oxygen for this to happen, so if your wine went bad, perhaps you had a leaky cork.

Imagine that your delicious bottle of wine indeed did turn into vinegar. This would be sad, but let's do it for the sake of science, and maybe you can use it to make a salad dressing when we're done. Vinegar is a powerful microbial agent. As little as 0.1% in a solution kills most microbes, so our bottle is still safe from other kinds of spoilage. Just as we did with the sugar to ethanol conversion in wine, we will be able to figure out just how acidic our vinegar gets. As it turns out, each molecule of ethanol produces one molecule of vinegar. From this we can figure out what the pH of the vinegar is. All we need to know is how easily the hydrogen ions on the vinegar dissociate into solution, because as you learned in chapter 1, this is what determines the hydrogen concentration, which in turn determines the pH. We won't bother to go through the calculations here, but if we did, you would see that we'd expect a pH a little over 2. And, the real pH of vinegar is usually exactly this. Some fruit vinegars have slightly higher pH, but for wine vinegar, assuming that all of the sugar turns into ethanol, and all of the ethanol then turns into acetic acid, a pH of around 2 is about what we would expect.

Waiting many years for your wine bottle to turn into vinegar by accident, as described in this sidebar, is not the best strategy for making vinegar. Take a look at the recipe in sidebar 6. The only ingredients are sugar, water, and some fruit. You mix them all together, cover with a cheesecloth, and store in a dark, cool place for many weeks, with intermittent sugar feedings every two weeks. ⚛

SIDEBAR 6: VINEGAR

Fruit Vinegar

Ingredients
1 cup sugar
4 cups water
2 cups fruit peels, pulp, or fresh fruit

| 2 cups | 1 cup | 4 cups |

Directions

1. In a large jar, dissolve the sugar in the water, then add the fruit.

2. Cover with a cheesecloth. Store in a warm place for 1 week.

3. Strain out the fruit and let the jar sit in a dark, cool place for an additional 5 to 7 weeks.

4. Strain the vinegar into bottles and store in the refrigerator. It will last for at least 4 months.

By now you know enough about fermentation to decipher the science of each of the steps in this recipe. For example, you know that the sugar will be turned into ethanol, which is then turned into acetic acid. Over time, the pH decreases, which means that fewer and fewer microbes can survive, and the microbes that are producing the acetic acid can thrive and take over the solution. You also know that covering the jar with cheesecloth, as opposed to some airtight material, ensures easy access to oxygen; we say that the chemical reaction requires aerobic conditions. An alternative recipe is to combine 1 cup of unpasteurized red wine vinegar or apple cider vinegar with 2 cups of wine or mead (you can use the mead recipe earlier in this chapter). Cover with cheesecloth and store in a cool, dark place for a few weeks, tasting often. The vinegar is done when the alcoholic flavor is gone, the smell is sharp and the taste right. ⚛

FLAVOR

If you think about what we did at the start of this chapter from a food safety perspective, what we accomplished is quite marvelous: thanks to the production of two tiny but highly specific molecules, alcohol and lactic acid, the yeast and bacteria were able to outcompete potentially harmful bacteria and the food is now safe to eat. Moreover, it will stay that way for quite some time. All this is due to the chemical reactions inside the microbes.

But other small molecules, in addition to the lactic acid and alcohol, have also been produced. These are the many molecules responsible for the vastly different flavors of fermented foods. Nowhere else in cooking do you produce flavors with such immense diversity. Just think about it. The list of foods with which we started this chapter included stinky cheeses, tangy pickles, buttery white wine, and deep and rich soy sauce. All of these foods owe their distinct flavors to microbes. Indeed, flavor production is one of the main reasons we continue to engage in this age-old practice.

Recall the browning reactions that occurred when we heated steak and sugar at hot temperatures. These were caused by complex breakdown and subsequent reactions of carbohydrates and proteins. The flavors of food fermentations are very similar in that they also come from large food components that have been broken down into smaller molecules. This is why you'll sometimes encounter flavor molecules in fermented foods that also appear in Maillard reactions. But food fermentations also have many, many other flavors. And with fermentation the flavor is not due to random reactions from heat, but rather are the products of enzymes as the microbes break down food molecules for energy. The flavor still depends on the specific makeup of the food—even small changes can impact flavor significantly—but the microbes do the work.

FLAVORS FROM PROTEINS AND FATS

All of the recipes so far in this chapter have involved some form of carbohydrate. We learned that microbes love sugar. But let's say you wanted to ferment fish or heavy cream, a common practice around the world. What would you eat if you were a microbe? Well, surely not carbohydrates, because there aren't any. Fish consists entirely of fat and protein, and cream is practically only fat in water, so you'd have to make do with this. And indeed, there are microbes that would be entirely happy with this menu. Lucky for us, proteins and fats have the potential to create a whole host of interesting flavors. The reason, of course, is that they are made up of atoms other than those in carbohydrates, which are mostly made of carbons, oxygens and hydrogens. Proteins, for example, contain sulfur and nitrogen, which can give rise to flavor molecules that are meaty and earthy. Some of them we've already encountered in the Maillard reactions, where proteins were also broken down. But perhaps the most distinct flavor molecules of all come not from sulfur and nitrogen but from the backbone of the protein chain itself. All proteins have this backbone, which is a string of repeating units with the same atoms. If you look closely and compare, you'll see that each unit looks very similar to molecules with names like cadaverine and putrecine. Yes, they smell exactly like their names suggest—hardly something you'd like in your food. Nevertheless, these molecules form very easily when proteins are fermented for a long time and they appear in foods around the world that are considered delicacies (although for the uninitiated they no doubt take some getting used to). Surströmming, the Scandinavian fermented herring, is *supposed* to have this flavor. You make it by leaving barrels of salted herring out in the sun for weeks while letting salt-loving bacteria break down the proteins. Another example is Icelandic fermented shark, which is produced with the same idea, although here you go as far as burying the fish in the ground, at least according to traditional recipes.

If you can't get your hands on these exotic foods but still want to experience the flavor profiles that result from fats and proteins, look no further than your closest cheese shop or supermarket cheese counter. The number of cheeses can be staggering. They range so widely in flavor that it's hard to imagine that they all start out as ordinary milk, but they do. And the intense flavors, especially in the cheeses that have been aged for long periods of time, owe their flavors to proteins and fats that have been broken down by complex successions of microbes.

There are many other kinds of protein fermentations that use a mix of traditional preservation techniques, such as smoking, drying, and curing. For example, salami, the Italian sausage, is produced by a combination of salt-curing, air-drying, and fermentation. Similarly, dashi, a traditional Japanese base for soups and stocks, contains the tuna-like fish bonito, which undergoes a long process of boiling, smoking, drying, and fermenting. The resulting product, called katsuobushi, has rich and complex aromas that make the base for a delicious soup.

FERMENTATION WITH MOLDS

In addition to yeast and bacteria, molds represent the final of the three most prevalent food fermentations. Unlike yeast and bacteria, which are single-celled microbes, molds consist of many cells and can grow into structures we can see with the naked eye. You know them as the fuzzy mold on old bread. Molds are especially important for fermentations of proteins and fats. In fact, they occur in many of the meat and dairy fermentations we just encountered. But where molds really shine is in the making of many traditional Asian foods like sake, miso, and soy sauce, to name a few. The star of them all is a mold named *Aspergillus oryzae*, which is often used in combination with soybeans or rice. Miso, for example, is made by first incubating boiled rice with *A. oryzae*, then adding the rice to soybeans, after which the fermentation continues. Since both rice and soybeans contain a fair amount of proteins and fat—close to 60% for soybeans—the flavors are rich and savory. Soy sauce starts out much the same and can be fermented for years to truly develop its deep flavors.

Are you ready to test your understanding of flavor engineering via fermentation? If so, here is a thought experiment: Imagine that you are a maker of sake, the famed Japanese rice wine. Sake is made by fermenting rice twice, first with *A. oryzae*, which breaks down the complex carbohydrates into simple sugars, and then with alcohol-producing yeast, which act on these sugars. If you are a distinguished sake maker, you would make sure to "polish" the rice grains before any of these steps, a process that involves grinding off the outer layer of the rice grains so that only the center is left behind. Usually as little as 35%–70% of the rice grain would remain. This seems like a huge waste. Why would you do all this extra work to get rid of most of your rice?

The answer lies in the anatomy of rice grains. The bran that surrounds the seed contains more protein and fat than the endosperm at the center that feeds the growing germ. For this reason, sakes that have not been polished at all tend to have hints of the savory, earthy flavors we expect from proteins and fats. This is fine in miso and soy sauce, but for sake we want light and floral, the kinds of aromas you get from carbohydrates. By polishing off the bran on the surface, only the starchy endosperm is left behind. Not surprisingly, sakes from polished rice are much more expensive than nonpolished ones. After all, you got rid of all that rice and have to make up the profits somehow. But this is not the only reason. Many people are willing to pay the higher price because the resulting sake can be so light, smooth, and delicate. Here's an experiment in flavor detection: try tasting sake with different degrees of milling next to each other. We recommend getting sakes at the extremes: futsushu sake, made from rice that has barely been polished at all, and daiginjo sake, made from rice that has been polished all the way down. A specialist in a liquor store should be able to help you find them. You should be able to taste the interesting effect of eliminating the proteins, a clear example of how small differences in ingredients can have big effects on flavor.

Chocolate and Coffee

The science of microbial cooking comes together beautifully in two of the perhaps most-loved fermented foods of all: chocolate and coffee. These are special fermentations in several ways. First, both are entirely "wild" fermentations in that they rely on the microbes in the environment and are not kick-started by adding specific cultures. This is not new—we saw that sauerkraut works the same way. What's different about coffee and chocolate though, is that they involve a mix of several types of microbes: yeast, lactic acid bacteria, acetic acid bacteria, molds, and more. These microbes occur in succession, which in chocolate can look something like the following: The pods, or fruits, of the cacao tree are harvested, then the cacao beans are removed from the pods and put in a box to ferment. The beans at this point are inedible and bitter, but the white mucilage that surrounds them is slightly sweet, making the environment perfect for ethanol-producing yeast. When the yeast growth diminishes, lactic acid bacteria are able to take over, followed by acetic acid bacteria, which thrive in the newly acidic environment. At the end come the molds. Each type of microbe thrives at its ideal growth conditions and changes the environment so that it appeals to the next succession of microbes. Together they transform the beans.

Coffee fermentation is similar. The main difference is that the mucilage around the smaller coffee beans is less sweet than that around the cacao seeds, so instead of yeast, the coffee fermentation is initially dominated by lactic acid bacteria, which quickly bring the pH down. The coffee fermentation is also shorter, a day or two, compared to about a week for cacao, but as with cacao it involves a range of microbes.

What's amazing about both of these very common fermentation reactions is that they are actually not very well known despite having long histories. We are just starting to understand them in detail. But we do know that both fermentations are predictable and reliable. Predictable because we get more or less the same succession of microbes each time, even with very limited human intervention. And reliable because the resulting products tend to be highly usable. While many other commercial fermentations have become highly controlled processes, where temperature and humidity are closely monitored and the desired microbes are purposely added, fermentation of cacao beans and coffee beans is

FIGURE 4 Fermenting cacao seeds

still done in the traditional way. Given that chocolate and coffee support billion-dollar industries, this is quite remarkable.

The next step after fermentation in both chocolate and coffee production is to allow the seeds to dry in the sun, which reduces the moisture content and stops the microbial activity. Next, the seeds are roasted. The fermentation reaction has produced ethanol, lactic acid, acetic acid, and more, and these molecules have diffused into the seeds and made flavor precursors. During roasting, Maillard reactions break down these molecules further, creating the intense flavors from coffee and chocolate that we know and love.

Fermentation in Haute Cuisine

The recipes for food fermentations have often remained the same for generations, if not millennia. They are robust. The same ingredients in combination with the appropriate microbes yield the same final food. Grapes and yeast make wine, cabbage and lactic acid bacteria gets you sauerkraut, aspergillus and soybeans produce miso, and so on. But what are the limits of these amazing little flavor machines known as microbes? Are there untried combinations of substrates and microbes that would result in novel flavors and foods?

These types of questions are currently being asked by chefs in many areas of haute cuisine. David Chang of New York's Momofuku wanted to know: What if you take the mold *Aspergillus oryzae* and, instead of feeding it soybeans as you do when making miso, feed it something else, like chickpeas or sunflower seeds. You would be pairing a microbe that traditionally renders flavors common in Asian food with a completely novel ingredient. Since chickpeas and sunflower seeds have a different composition of ingredients, you get different flavors, reminiscent of miso, but nonetheless novel. David Chang found that it worked and called his new creation "hozon."

The team at restaurant Noma in Copenhagen took this idea further by producing a range of misos including peaso, breadso, maizo, and hazelnut miso. See the recipe for their famous "peaso" in the sidebar. Try it at home and see for yourself how the flavors work out.

SIDEBAR 7: PEASO

Peaso

Ingredients
800 g dried yellow split peas
1 kg fresh barley or rice koji (recipe follows)
100 g noniodized salt, plus extra for sprinkling

Directions
1. Put the dried peas in a large container and cover them with double their volume of cold water. Set aside for 4 hours at room temperature.

2. Drain the peas, put them in a large pot, and cover them again with double their volume of cold water. Bring the water to a boil, then reduce the heat to a bare simmer, skimming away any starchy foam that rises to the surface. Cook for 45 to 60 minutes, stirring every 10 minutes or so, until the peas are soft enough to crush between your thumb and forefinger without applying much pressure.

3. Drain the peas, spread them on a rimmed baking sheet, and cool to room temperature. Weigh the cooled peas; you should have close to 1.5 kg, but the amount of water the peas absorb in the soaking and cooking process will always vary. (If you have more than 1.5 kg, you can set the extra peas aside for another use. If you have less than 1.5 kg, you'll simply need to adjust the ratio of the other ingredients. The amount of koji needed is 66.6% of the weight of the cooked peas; the salt is 6.6%. If, for instance, you wind up with 1.3 kg of peas, reduce the amount of koji from 1 kg to 866 grams and the salt from 100 grams to 86 grams. These are exact ratios that you should adhere to if you want your peaso to turn out the way it's intended.)

4. Wearing latex or nitrile gloves, put the cooked peas in the hopper of a sanitized meat grinder and grind them with a medium die into a very large bowl or container. Next, grind the koji and add it to the peas. (Alternatively, you can use a food processor, but be careful that you don't overprocess things. You're not aiming for a purée—a coarse meal will suffice. As a last resort, if you don't have a meat grinder or food processor, you can mash the peas in a large mortar and pestle and crumble the koji by gloved hand.)

5. Mix the ground peas and koji with a spoon, then squeeze a small handful of the mixture to check the texture and moisture content:

    If it easily forms a compact ball, you're good to go.

    If the mixture crumbles, it's too dry and you'll need to add some water to hydrate it. However, it's vital that you maintain the 4% salt ratio, so any liquid you add to the mixture should have the same salt content. Make a quick 4% salt brine by blending 4 grams salt into 100 grams water with a handheld blender or whisk until the salt has completely dissolved. Add a little bit at a time to the pea mixture until you've achieved the proper texture.

    If the mixture oozes out of your hand when squeezed, it's too wet; the peas may have been overcooked or improperly drained. Too wet is more difficult to correct than too dry, but not impossible. To correct in such an instance, spread the mixture on a parchment-lined rimmed baking sheet in a thin, even layer and dry in an oven or a dehydrator at 104°F (40°C), giving it the squeeze test frequently, until it reaches the texture you're looking for.

6. Add the salt and mix thoroughly.

7. Working with one gloved handful at a time, transfer the peaso to a 5 L nonreactive fermentation vessel (glass, plastic, ceramic, or untreated wood) and pack it as tightly as possible. Start at the edges of the container, forcing any air out, then work your way toward the center. Punch the mixture down with your fists after each addition to ensure that it's well packed. Smooth and flatten the top of the peaso and lightly sprinkle the surface with salt to help prevent mold from forming. Place a sheet of plastic wrap over the top, in direct contact with the peaso, making sure it reaches all the way to the edges. Finally, wipe down the walls of the container with a clean paper towel.

8. Now you'll need to weigh down the peaso. As the peaso ferments and yields tamari, the weight will keep the mixture submerged in liquid in the same way that a lacto-ferment like sauerkraut rests beneath its juices. If you like, you can purchase specially designed fermentation weights online that will fit the circumference of your

fermentation vessel. Otherwise, the simplest method is to use a flat dinner plate that fits inside your fermentation vessel snugly. If you're using a plate, bear in mind that the plate will sink over time and that eventually you'll need to remove it, so be sure it doesn't fit *too* snugly or you won't be able to take it out. Place the plate right-side up on top of the peaso and press it down with your hand. Now procure a rock, a brick, or a few cans that weigh roughly half as much as the peaso—about 1.5 kilograms. Place the weights in plastic bags to keep things sanitary and distribute them evenly over the plate.

9. Your peaso will do fine fermenting on your kitchen counter at room temperature, but Noma ages their peaso in a dedicated room held at 82°F (28°C) for about 3 months. It should age well in either scenario, although an extra month might be necessary at room temperature, and you can certainly leave it even longer than that, if you like. The peaso gets much richer, with darker, earthier tones, the longer it ferments.

10. Check the peaso's progress after 3 or 4 days. It won't look all that different from when you started. If anything, it will be slightly more aromatic. If that's the case, it's going well. If you notice it souring like a lactic ferment, with a lot of tamari pooling on top, it means your mixture was too wet and you'll have to start over. If you're fermenting in a clear container, you may notice small air pockets forming throughout the peaso. These are part of the normal fermentation process; they will subside over time.

11. After the first couple of weeks, open the peaso every week or two to check its progress. Be sure to wear gloves to avoid introducing contaminants. At some point, you may find white mold growing on the surface. That's totally fine. In our experience, it's usually a patch of koji that has managed to grab a foothold on an exposed portion of the mixture. But even if it's another mold, if the miso is packed tightly, the mold won't be able to penetrate the surface. When you need to taste the peaso, scrape a bit of the mold to the side to get underneath, but don't remove it completely until you harvest the whole batch, lest more come back in its place.

12. The peaso is finished when the texture has softened significantly, the taste of salt has subsided slightly, and all manner of sweet, nutty tones have emerged—usually somewhere between 3 and 4 months. It should have a mild acidity without being overly sour. The texture of the peaso will be slightly nubby, so if you'd prefer a very smooth paste, blend the peaso in a food processor—add a bit of water to help it spin, if necessary—after which you can pass it through a tamis if you're looking for a truly velvety texture.

13. You can pack the peaso into airtight jars or containers and store them in the fridge for use within the month. Any longer than that, store it in the freezer to keep its flavor freshest; just pull it out as you need it.

Koji

Ingredients
500 g pearl barley or short-grain rice
1 teaspoon powdered koji spores (*Aspergillus oryzae*)

Directions
1. Rinse the barley thoroughly under cold water, then soak for 4 hours. Drain.

2. Steam the barley until tender but not falling apart, 20 to 30 minutes. You can use a regular steamer, or a sieve or colander set in a pot with a lid.

3. Add the cooked barley to a tray or rimmed baking sheet that has been lined with a towel (clean, sanitized with steam, and wrung dry). Break up the barley to avoid clumping. Let cool to 86°F (30°C).

4. Inoculate the barley with the koji spores by sifting it through a tea strainer over the barley. Wearing gloves, mix thoroughly.

5. Cover with a damp kitchen towel and ferment at 86°F (30°C) and a humidity of 70%–75%.

6. After 24 hours, you'll begin to see mold growth. Mix and break up any clumps.

7. After 48 hours, the koji will be fully grown. It will be completely covered in a light green or white fuzz and have a fruity smell. Cool in the refrigerator, then store in the refrigerator or freezer. ⚛

Perhaps one of the most striking examples of fermentation in haute cuisine is the work being done by the team at restaurant Mugaritz in Spain's Basque Country. We met them before in the chapter on enzymes where they used pectinase to "cook" an apple. Here, they add *Rhizopus oryzae*, a fungus usually used for making the Southeast Asian meat-like soy product called tempeh, to the surface of a small crab apple in a completely novel application. The fungus grows a fine mold that covers the apple; only the stem sticking up reveals what it is. Despite the association with spoiled food—as a diner your first reaction as the dish is brought out is slight shock—the apple looks beautiful and tastes delicious, an artful play on the fine line between rotting and fermenting.

SIDEBAR 8: MUGARITZ RHIZOPUS APPLE

Mugaritz Rhizopus Apple

Ingredients
1 g ascorbic acid

1 L water

4 Txalaka (wild Basque apples) or Granny
  Smith apples

Tempeh starter (*Rhizopus oryzae*)

Vodka

Directions
1. Dissolve the ascorbic acid in the water.

2. Peel the apples very gently, trying to keep their natural shape. As you peel each apple, submerge it in the ascorbic acid mixture to prevent enzymatic browning. (If you are only making a small number of apples, you can ignore this step.)

3. Bring a pot of fresh water to 194°F (90°C). Add the apples and let them heat for 2 minutes.

4. Remove the apples and let them cool.

5. Sprinkle the tempeh starter over the apples until they are lightly covered.

6. Place the apples on a tray, cover with plastic wrap, and let ferment at 86°F–95°F (30°C–35°C) for 24 hours.

7. Place each fermented apple on a plate.

8. Add a drop of vodka. ⚛

Some of these foods may seem like extravagant examples, but you can enjoy the craft of skilled fermentation practitioners at local breweries, cheese shops, or even bakeries. And it's not just a select few experimental restaurants that are harvesting the creative powers of microbes. The next time you go out to eat, keep your eyes out for fermented ingredients on the menu. They are increasingly sneaking their way into many types of restaurants, in various shapes and forms. Perhaps by now they have also snuck their way into your kitchen? If you started one of the fermentation recipes at the beginning of this chapter, they should soon be ready to eat. Congratulations! We hope you feel inspired to explore further. As you do, we hope you enjoy the bubbly and smelly jars, all signs of marvelous scientific processes carried out by millions of little microbes that are doing the cooking for you.

Conclusion

Throughout this book, we have ingested a multicourse dinner of sorts, where the individual dishes have been scientific explorations into all types of foods. We have now arrived at the end, so it is fitting to have dessert. And what could be better to round off this meal than chocolate? Not only is it a delicious food, but the making of chocolate involves essentially all of the topics in this book. So, please, have a piece of chocolate, and while you let it slowly melt in your mouth, let's reminisce together about the meal we just shared.

This entire book could have been explained through the single lens of chocolate and the many different ways of preparing it. Now that you have read the book, we can explain why. In chapter 1 we introduced the idea that food consists of both flavor molecules and texture molecules. For chocolate, these molecules come from cacao beans. The flavor molecules are produced when the beans are fermented and roasted, the latter of which produces Maillard reactions. We discussed these two mechanisms in chapters 7 and 2, respectively. In chapter 7 we also learned that what's ultimately fueling the fermentation is not the microbes themselves but the enzymes within the microbes. We discussed how enzymes work in chapter 3. The concept of diffusion, which we discussed in chapter 4, is critical for understanding how the molecules that are produced by the microbes as they ferment the mucilage surrounding the beans (lactic acid, ethanol, and so on) diffuse into the cacao beans and transform them. And the idea of protein denaturation from chapter 2 helps us understand how we can control the fermentation itself, heat being the entity that would not only disrupt the flavor-producing enzymes inside the microbes, but eventually also render them nonfunctional to the point of killing the microbes themselves.

Moving on from the flavor molecules to chocolate's texture molecules brings us further into the book. In chocolate, the texture molecules consist almost entirely of fat molecules that have been extracted from the milled cacao beans following fermentation and roasting. When the chocolate is warm, it is a liquid whose viscous properties can be understood per our discussion in chapter 5. More specifically, the liquid is in fact a colloidal suspension—it consists of small suspended cacao particles—and, if the chocolate is in the form of milk chocolate, the liquid is also an emulsion (recall chapter 6). Moving on, when the heated liquid is chilled, the fat molecules arrange themselves into specific structures, and the liquid undergoes a phase transition and becomes a solid (see the sidebar in chapter 2). The solid chocolate bar, like all solid foods, has an elastic modulus, which is important because we want our chocolate to have a certain snap (chapter 5).

Phew. In telling the story of how chocolate is made, we really did touch on almost every topic in this book! And this is not unique to chocolate. As you start to consider the foods around you more closely, you'll see that any one food or recipe often combines numerous scientific principles. We hope this book will allow you to continue discovering them. But even more than this, we hope you will continue asking questions about the foods and recipes around you just as we have done in this book.

In our course, the idea of asking questions plays an important role. At the end of the semester, students spend several weeks working on their own projects. Just as professional chefs are innovative experimentalists, we want students to apply what they have learned by asking their own questions and finding their own answers. The first and hardest part is choosing the right question: it should be interesting enough so that we should care about the answer, but we also need a clearly laid out plan for how we are going to get to an answer. Armed with a question, students use their culinary skills together with scientific knowledge to get to an answer and hopefully be rewarded with a recipe that is as delicious as it is insightful. Of course, much of the time, such plans end in failure; we then have to take what we have learned from our failed experiment and figure out what we can do with it, and construct a new plan. Such failure occurs constantly for working scientists and chefs who are trying to be innovative. Persistence however can pay off: by iterating on ideas, we can get to something that works.

We'd like to end this book by encouraging you to do the same. What questions do you have about the foods and recipes around you? Simply opening a cookbook and asking why a recipe works can generate tons of questions. What experiments can you do to find answers to your questions? Be a scientist in your own kitchen. We will be cheering you on from afar.

And why stop at cooking? The benefits of curiosity and asking questions extend much beyond cooking and can go a long way toward helping us understand, and improve, the world we live in.

Just as eating is not an activity aimed at nourishment alone, but also holds important cultural and social dimensions, we hope the contents of this book will nourish conversations and connections over many dinner parties, cooking endeavors, and social gatherings ahead.

As a parting note, we'd like to share a drink recipe from food scientist Dave Arnold. Enjoy!

SIDEBAR 1: DAVE ARNOLD'S THAI BASIL DAIQUIRI

Dave Arnold's Thai Basil Daiquiri

Ingredients

5 g large Thai basil leaves (about 7)
2 ounces (60 mL) Flor de Caña white rum
   (40% alcohol by volume) or other clean white rum
¾ ounce (22.5 mL) freshly strained lime juice
Scant ⅝ ounce (18.75 mL) simple syrup
2 drops saline solution or pinch of salt

Directions

1. Nitro-muddle the Thai basil in a shaking tin.*

2. Add the rum and stir.

3. Add the lime juice, simple syrup, and saline solution or salt.

4. Check to make sure the mixture isn't freezing cold.

5. Shake with ice and strain through a tea strainer into a chilled coupe glass.

Add a small amount of liquid nitrogen to the leaves already in the shaking tin and swirl by holding the tin near the top, far from the dangerously cold bottom. When the boiling calms down, you want just a few millimeters of liquid nitrogen on the bottom. Crush with a wooden or plastic muddler. The frozen leaves are brittle and easily break into tiny pieces—this adds maximum fresh flavor to the drink. Normally, broken leaves activate enzymes that lead to browning and an oxidized taste (recall our discussion of pesto in chapter 3). The low temperature of the liquid nitrogen slows down the enzymes, and the leaves don't thaw until they are submerged in rum in step 2, where the high alcohol content continues to keep the enzymes turned off. If you don't have liquid nitrogen, you can also blend the basil with gin in a blender and then strain it out; this works almost as well. These techniques result in a remarkable green colored drink with the freshest basil flavor you can imagine. ⚛

Acknowledgments

An endeavor of this magnitude would only happen with help, support, and work from many people. For us this book is more than a book. It is the culmination of more than a decade of work, spurred on by the dream of transforming science education by teaching science through cooking. In acknowledging the contributions to this book, we must first and foremost thank the people who helped us build up the class at Harvard with the same name. The book is our attempt to write down our journey on paper—but the ways of teaching that we outline here have been developed over the years in close collaboration with the students at Harvard College and the graduate teaching fellows who helped guide them. The critical feedback of the students honed our explanations significantly and has helped us learn how to present the scientific method in the context of haute cuisine. Harvard graduate students have had an enormous impact on the material and its mode of presentation.

Before thanking individuals, indulge us with a brief origins story:

We launched Science and Cooking in the fall of 2010. For a time, it was the most popular class on Harvard's campus. The first day of the first class, a beautiful fall day in September 2010, the building at Harvard where science lectures took place was completely mobbed, with students densely packing the ground floor, clamoring to get a place in the 350-seat classroom where the lecture took place. Brenner himself was rushing to the class to give the first lecture, and had to push his way through the crowd to get into the lecture hall. Hundreds upon hundreds of students couldn't get into the room. After the fact we learned that some of the students managed to get a seat for the first lecture by sitting through the introductory physics class that was given just before. The first year enrollment was limited to 300 seats, driven by the constraints of the laboratory. Over 700 students entered a random lottery for these spots. We explained to the students that the probability of gaining admission to the class was much higher than the probability of gaining admission to Harvard, but astronomically higher than getting a reservation at Ferran Adrià's restaurant elBulli, where there were 8 million reservation requests for 2,000 slots.

The class came about through a series of lucky coincidences that are worth recounting for posterity. It all started when a (then) postdoctoral fellow named Otger Campas proposed to Brenner and Weitz that we invite Ferran Adrià to come to Harvard to give a lecture. Neither Brenner nor Weitz knew much about food or cooking, but Campas explained that Adrià's creations used principles from the materials research in which many people at Harvard were engaged. It was a brilliant idea, presaging all that followed. So Adrià came to give a special lecture on "Cooking and Science." The lecture was sponsored in part by the science outreach section of a National Science Foundation funded Materials Research Center at Harvard. The Center has sponsored outreach activities since its founding in the 1960s, and continuing on to this day. Adrià's lecture was open to the public in Cambridge, MA, and began at 6:30 pm. The room for the lecture had 250 seats and started filling up at 3:30. By 4:30 it was packed and we scurried to get overflow rooms. Even still there were hundreds of Cambridge residents who couldn't get into the lecture, a sign of things to come. Luckily, Brenner didn't go to the lecture and instead went to a faculty meeting, where the faculty was encouraged to develop new classes for Harvard's newly developed Program for General Education. This was a revitalization of the core curriculum for undergraduates and aimed to connect the experiences of students in the classroom to those in the world. The next morning Brenner, Campas, and Weitz met with Adrià and his righthand person, Pere Castells, to discuss ways of working together. The perfect synergy was at hand: we would launch a class to teach students about science through haute cuisine. Adrià and Castells proposed recruiting a star-studded cast of chefs to help teach the class. José Andrés and his ThinkFoodGroup was an early proponent and advocate for the class through his encouragement, his substantial energy, and his early financial support.

That the class was allowed to take place in the first place was itself a bit of a miracle. In 2010, the General Education program at Harvard College was new, and promised a new way of teaching. Intellectual rigor and connection to the world around us were paramount. When we proposed this class, we did not have the detailed syllabi, problem sets, and learning goals that classes are usually required to produce. Our proposed content was little more than a promise and a star-studded cast of chefs for the first year. To this day, we are unsure why Harvard decided to have faith in us and let us go forward. We are very grateful to Stephanie Kenen, then the director of the program for General Education, and Jay Harris, then the dean for Undergraduate Education, for their trust. Stephanie was the source of countless pieces of wisdom and advice both in getting the class started and, over the years, helping us surmount the challenges along the way.

The first year of the class was preceded by a summer of intense curriculum development and dinner parties. This was led by the indefatigable Dr. Amy Rowat, who delayed the start of her own faculty career at UCLA to help launch this class. Within a few short months we had to construct a syllabus with problem sets, laboratory exercises, lectures, and exams that were rigorous enough to pass muster with our Harvard colleagues but at the same time kept the fun and wonder of cooking. Amy was assisted by a team of graduate students: John McGee, Jennifer Hou, Emily Russell, Larissa Zhou, Naveen Sinha, Ben Miller, and Sam Lipooff. The then executive chef for the Boston Red Sox, Steve (Nookie) Postal, participated in our early brainstorming sessions with great consequence. For example, during an enthusiastic discussion of the physics of cooking steaks, he memorably told one of us, "You don't know how to cook a steak! I know how to cook steaks."

Over the years, we have been assisted in teaching the class by more than a hundred graduate students, too numerous to mention here. In their stead, we thank the head teaching fellows for each year of the class, who had the job of managing the extreme production that the class had become (Naveen Sinha, John McGee, Aileen Li, Mary Wahl, Marina Santiago, Katherine Phillips, Mishu Duduta, Laura Doherty, Anjali Tripathi, Zach Gault). We also thank the lab managers who have managed the not insignificant task of overseeing the labs and lecture demonstrations for one of the most organizationally complex courses on Harvard's campus (Héloïse Vilaseca, Helen Wu, Sameer Tyagi, Denise Alfonso, Mai Nguyen, and Patricia Jurado Gonzalez). The administrative requirements of this class, especially at its inception, were significant. Not only did we have to arrange schedules for world class chefs and their entourages each week of the class, organize a massively popular public lecture series with advertising, sponsor outreach, order food for 300 students each week of the lab, organize a science fair with sponsors and students alike in an iconic Harvard dining hall, and so on, Christina Andujar did it all with panache and a smile and truly was a backbone upon which this course was built. Since 2014, these tasks have been managed with unflinchingly good nature, energy, and skill by a collection of individuals, most notably and for several years, the amazing Mai Nguyen, Dawn Miller, and Patricia Jurado Gonzalez.

A major inspiration for the class was Harold McGee, the famed author of the classic book *On Food and Cooking*. We were very fortunate that Harold agreed to participate in the class from the very beginning, and has served as a mentor and a sounding board to us ever since. The first public lecture featured Harold, Ferran Adrià, and José Andrés. Harold is the voice of wisdom, explaining the history of science and cooking and knowing more than anyone else we have met about how the subject has developed. Every year since, Harold has come to Harvard to lecture in the class, and his lectures continue to be a true highlight. In our online HarvardX class, every week of the class has reflections by Harold, whose eloquence and wisdom are second to none.

We owe a special thank you to Daniel Rosenberg for being an infinite source of inspiring lecture demonstrations that influenced the content of the class. Many of the most iconic science and cooking demonstrations originated with Daniel, such as the making of liquid nitrogen snow to demonstrate phase transitions, the ricotta cheese on a microscope slide to illustrate diffusion, and many, many more.

We would like to thank the administration of Harvard University, particularly the then dean of the School of Engineering and Applied Sciences, Cherry Murray, and the dean of the faculty, Michael Smith, for letting this course happen and for spurring us along. Dean Frank Doyle and Executive Dean Fawwaz Habbal have continued to provide important support and encouragement. Through the students and administration, Harvard gave us the freedom to try something that was both out of the ordinary and loads of fun.

Finally, we want to thank the many chefs who taught in this class and inspired us by their examples and their creations. One of the biggest surprises of this class to us is how similar world class chefs are by personality and temperament to world class scientists. Both are conditioned to try things that don't work, to experience failure on a regular basis, and to learn how to pick themselves up off the floor (having failed miserably, albeit with passion and energy) and turn the failure into something that is positive and interesting. This class certainly is an example of a successful failure—and we are grateful to the chefs who worked with us for their energy and effort in bringing this vision to creation. We would especially like to thank Pere Castells, of Ferran Adrià's elBullifoundation, who has been at our side from the very beginning, always making intelligent suggestions and proposals for chefs, for recipes, for scientific concepts, and for how to present this subject with the class that it deserves. His energy for this subject has been boundless, starting with his first enthusiastic explanations of his discovery of inverse spherification. A special thank you to Ferran Adrià, Jose Andrés, Harold McGee, Bill Yosses, Joanne Chang, and Dave Arnold, who have played special roles as patrons and regular visitors to our class.

In addition, and roughly by first year of appearance, we would like to thank all of the chefs who have participated. They have inspired us and taught us more than we can explain. Joan, Jordi, and Josep Roca (El Celler de Can Roca), Wylie Dufresne (wd~50, Du's Donuts), Grant Achatz (Alinea), Dan Barber (Blue Hill

Farms), Carme Ruscalleda (Sant Pau), Nandu Jubany (Can Jubany), Carles Tejedor (Via Veneto, Oil lab), Enric Rovira (Master Chocolatier), David Chang (Momofuku), Ramon Morató (Master Chocolatier, Barry Callebaut), Nathan Myhrvold (Modernist Cuisine), Fina Puigdevall (Les Cols), Pere Planagumà (Les Cols), Carles Gaig (Fonda Gaig), Paco Perez (Miramar), Raül Balam Ruscalleda (Moments), Jack Bishop and Dan Souza (America's Test Kitchen), Ted Russin (Culinary Institute of America), Mark Ladner (Del Posto), Steve Howell (NASA), Dominique Crenn (Atelier Crenn), Martin Breslin (Harvard University Dining Services), Christina Tosi (Milk Bar), Daniel Humm (Eleven Madison Park), Jody Adams (Rialto, Trade, Saloniki Greek), Jim Lahey (Sullivan Street Bakery), Andoni Aduriz and Ramon Perisé (Mugaritz), Bryan and Michael Voltaggio (Volt, Ink), Tara Whitsitt (Fermentation on Wheels), Tom Colicchio and Gail Simmons (*Top Chef*), Mei Lin (*Top Chef* season 12 winner), Spike Gjerde and Lauren Sandler (Woodberry Kitchen), Mark Post (Maastricht University), Virgilio and Malena Martínez (Central), Margarita Forés (Cibo restaurants), Vayu Maini Rekdal (Young Chefs Program), Heloise Vilaseca (El Celler de Can Roca), Ángel León (Restaurant Aponiente), Sister Noella Marcellino (Abbey of Regina Laudis), Mateo Kehler (Jasper Hill Farm and Caves), Mario Batali (Babbo), Lidia Bastianich (Lidia's Kitchen, Eataly), Sandor Katz (*The Art of Fermentation*), Corey Lee (Benu), Roberto Flore (Nordic Food Lab), Gabriel Bremer (La Bodega, Salts), Michael Harlan Turkell (Acid Trip), Ayr Muir (Clover Food Lab), Vicky Lau (Tate Dining Room & Bar), Joxe Mari Aizega, Juan Carlos Arboleya, and Diego Prado (Basque Culinary Center), Tiffani Faison (Sweet Cheeks BBQ, Tiger Mama), Massimo Bottura (Osteria Francescana), David Zilber and Jason White (Noma), Janice Wong (2am:dessertbar), Nick Digiovanni'19 (Science and Cooking alum and *Master Chef* finalist 2019), Freddie Bitsoie (Smithsonian's National Museum of the American Indian), Marsia Taha (Gustu), and Selassie Atadika (Midunu).

We would also like to thank the team at HarvardX who were amazing in turning our course into an online class. In particular, the faculty director of HarvardX, Rob Lue, Director Annie Valva, Sarah Jessop, Casey Roehrig, Heather

Sternshein, Anna Trandafir, Marlon Kuzmick, Matthew Thomas, Alex Auriema, and the many additional team members who helped in countless ways.

Also, a special thank you to Victoria Shen and the team at Harvard Media and Technology Services, who have managed the video recordings of our lectures. If you have watched the YouTube lectures of the chef's presentations online, you have them to thank for the high quality.

The following individuals have all been critical to various aspects of the course: Karen Galvez and Douglas Woodhouse of the General Education Program, as well as Kate Zirpolo-Flynn, Arlene Stevens, and Matthew Zahnzinger of the School of Engineering and Applied Sciences for their organization and unfailing support with everything from running the class-wide admissions lottery at the start of the semester to arranging travel and accommodation for the visiting speakers. Also, Tim Roth for his assistance teaching online and hybrid versions of the class. Nabila Rodriguez Valeron, Nick Digiovanni, and Tim Roth for their help organizing the public lectures. Tracy Chang at Pagu for generously helping us locate rare ingredients and equipment from her own or her colleagues' local restaurants. Katherine Guenthner for designing beautiful posters advertising our lectures every year.

Over the years we have been lucky to receive generous sponsorship from a number of sources. Most importantly, a special thank you to the National Science Foundation, who supported us through both Harvard's Materials Research Science and Engineering Center (MRSEC), and the division of Mathematical Sciences. The NSF sponsored Adrià's very first public lecture that got us started. Moreover, their support and encouragement for developing substantive outreach activities that explain our scientific research to the public has been a continuous inspiration. We are also grateful to ThinkFoodGroup, Alicia Foundation, elBullifoundation, Wholefoods Market River Street, Fusionchef by Julabo, Lexus, Mont-Ferrant, Xertoli, Gastronomy Solutions, Escata, and 1933 cocktails for their sponsorship.

In the writing of this book, we are grateful to Helen Wu and Evan Liu for contributing early draft material for some portions of the book. We also thank

Patricia Jurado Gonzalez, Arvind Srinivasan, Jonah Brenner, Jenny Colwell, Alan Brenner, and Ronni Brenner for reading parts or all of the draft. Most importantly, a thank you to the amazing team at Norton, especially our editor, Quynh Do, for believing in the idea and guiding us in the process of writing this book.

Finally, we owe a special debt of gratitude to our partners, families, and friends, who have patiently supported us in, and been integral parts of, this decade-long adventure of science and cooking.

Photo and Recipe Credits

	Page 150	Photo by Amanda Justice
	Sidebar 2	Recipe and image courtesy of El Celler de Can Roca
	Figures 6 and 7	Images drawn from Cook My Meat, an app developed in collaboration with Professors Rob Miller and Fredo Durand at MIT and their students Kate Roe, Laura Breiman, and Marissa Stephens. (http://up.csail.mit.edu/science-of-cooking)
	Sidebar 4	Recipe courtesy of Central Restaurante Photo by César del Río
Chapter 5	Sidebar 1	Recipe and image courtesy of El Celler de Can Roca
	Sidebar 8	Recipe and images courtesy of Nandu Jubany s.c.
	Sidebar 9	Recipe courtesy of minibar by José Andrés Photo by © Peter Frank Edwards
	Sidebar 11	Recipe courtesy of Benu
	Sidebar 12	Recipe courtesy of Ferran Adrià Photo by © F. Guillamet
	Sidebar 13	Recipe and image courtesy of Carme Ruscalleda
	Sidebar 14	Recipe courtesy of William Yosses
Chapter 6	Figure 3	Photo by Arvind Srinivasan
	Sidebar 3	Recipe and image courtesy of Nandu Jubany s.c.
	Sidebar 4	Recipe and image courtesy of Nandu Jubany s.c.
	Sidebar 5	Recipe and image courtesy of El Celler de Can Roca
Chapter 7	Figures 1 and 4	Photos by Scott Chimileski
	Sidebar 7	Recipe courtesy of Noma and Artisan Workman Publishing Photo by © Evan Sung
	Sidebar 8	Recipe courtesy of Mugaritz Photo by Jose Luis Lopez de Zubiria
Conclusion	Sidebar 1	Recipe courtesy of Dave Arnold Photo by Travis Huggett

Index

Note: Page numbers in *italics* refer to illustrations or tables.

diffusion (*continued*)

heat diffusion in steak, 123–24, 132, 136–38, 145, 148–49

heat in molten chocolate cake, 129, 131–32

hydrogen ion diffusion in ceviche, 132–33

Layers of the Amazon, 155–58

to the naked eye, 124–26

overview, 123–24

as "random walk," 126–28, *134*, 186

in ricotta cheese production, 124–26

spherification and, 133–36

dispersed phase

colloidal suspension, 236

emulsions, 213, 214, 215, 222, 225, 236

foams, 228, 229, 234

Dufresne, Wylie, 118, 119, 122

Dumpling Wrappers, 199

eggs

century eggs, 86, 87–88, 91, 93, 96, 103

chicken serum albumin, *98*

cooking of, 76–80, 81–82

cooking temperature and, *39*, *77*, 79–80

gelation transition, 82

as heat-stable gel, 204–5

ovalbumin (egg albumin), 22, *23*, *75*, 96, 228

ovotransferrin, 80, 81–82

pasteurization, 76

phase transformations in cooking, 38–39

poached eggs, 101–2

sous vide cooking, *77*, 78–80

surfactants in, 219, 222, 225

Thousand-Year-Old Quail Eggs, 86, *87*, 90

transition temperatures in cooking, *39*

Einstein, Albert, 127–28, 145

elasticity

effect of polymers, 166

gels and, 197–98

overview, 165, 193

packing and, 196

polymer networks and, 196–98

volume fraction and, 195–96

ways to increase, 195–97

see also plasticity

elastic modulus (Young's modulus)

calculation of, 194–95

of common foods, *193*

equation, 195, 197–98

overview, 165, 193

of steak, 194–95, 206

see also solids

elBulli, ix–x, xiv, xvi, 54, 228

elBullifoundation, xi

El Celler de Can Roca, 26, 54

electric charge

charged amino acids, 86, 95–96, *97*

negative charges on casein proteins, 113

pH-dependent protein unfolding and, 95–99

transformations with, 85–86

Eleven Madison Park, 18

emulsifiers. *see* surfactants

emulsions

coalescence of droplets, 215, *216*, *217*, 221

continuous phase, *213*, 214, 218, 221, 225, 227, 236

cookie dough, 214, 227, 236

defined, 213

dispersed phase, 213, 214, 215, 222, 225, 236

gel stabilization of, 221

hollandaise sauce, 182, 183–84, 213, 222–23, 224

ice cream, 236–37

mayonnaise, 213, *214*, 223–24

oil-in-water emulsions, *213*, 218

opaqueness, 214

phase inversions, 222, 227

protein stabilization of, 220, 222, 228

separation of, 214, 215, *216*, *217*

solidification for stabilization, 221

starch stabilization of, 220

flavor (*continued*)

    sensory experience, 8–9

    separating smell from taste, 14

    taste molecules, 12, *13*, 14, 153–54, 230

flour

    gluten protein in, 22, *23*, 206

    packing, 172–73

    starches, *11*, 22

    water in, 11

    weight of one cup of, 172, *173*

Flour Bakery + Cafe, 43

foams

    angel food cake, 236

    continuous phase, 234, 236

    cookie dough, 227, 236

    culinary foam invention, 228

    dispersed phase, 228, 229, 234

    drainage, 228, 234–35

    egg whites, 227–28

    head on beer, 234–35

    ice cream, 230, 236–37

    overview, 227–28, 229–30, 234–35

    Oyster Foam, 229

    protein stabilization of, 228

    solids, 224, 228, 234–35, 236, 237

    whipped cream, 229, 235

    see also surfactants

food safety and temperature, 74

Forés, Margarita, 94

freezing point depression, 70–73

Fresh Mint Oil, 109–10

fried ice cream, 149–52

Fruit Vinegar, 257–58

fungi, fermentation with, 261–62, 263, 264, 268–70

Garlic Alioli (aioli), 225, 226

Garlic Oil, 203–4

garlic, surfactants in, 219, 225

gastric fluid, 18

gelatin

    from collagen, 185, 186, 205–6

    as hydrocolloid, 186, 198

    Jell-O, 193, 197, 198, 201, 206

    polymer networks (gels), 185, 197, 198, 201

    sheets, 156, 163, 190, 232, 234

    viscosity increased by, 185, 186

gelation, 82, 133, 221

gels

    agar-agar, 201

    Bergamot Gel, 56–57

    Black Truffle Gel, 203

    eggs as heat-stable gel, 204–5

    elasticity and, 197–98

    emulsion stabilization by, 221

    from hydrocolloids, 198

    Jell-O, 193, 197, 198, 201, 206

    Lime Gel, 232

    Lobster Filling, 200

    pasta as, 84

    Sherry Fluid Gel, 191

    spherification and, 107, 111, 134–35, 136, 165–66, 201

    Warm Codfish Gel, 202–3

    see also gelatin; solids

Giner, Joan, 69

Giner system, 69

gliadin, *23*, 206

gluconobacter, 257

gluten, 22, *23*, 206–8

gluten-free dough, 207

glutenin, *23*, 206

Grape Juice, 191

Grape Spheres, 190–91

growth curves, *250*, 251

Gymnema sylvestre powder, 160

heat

    calories and, 37–38

    defined, 37

About the Authors

Michael P. Brenner is the Michael T. Cronin Professor of Applied Mathematics and Applied Physics and Professor of Physics at the John A. Paulson School of Engineering and Applied Sciences at Harvard University. His research uses mathematics to examine a wide variety of problems in science and engineering, ranging from understanding the shapes of bird beaks, whale flippers, and fungal spores to finding the principles for designing materials that can assemble themselves, and answering ordinary questions about daily life, such as why a droplet of fluid splashes when it collides with a solid surface.

Pia M. Sörensen is Senior Preceptor in Chemical Engineering and Applied Materials at the John A. Paulson School of Engineering and Applied Sciences at Harvard University. Her research interests range from the science and engineering of food— including the chemical and microbial processes of food fermentations, the science of flavor, and scientific approaches to understanding the ancient history of food—to science education—the latter with an emphasis on online education and creative ways of teaching science in a liberal arts setting.

David A. Weitz is Professor of Physics and Applied Physics at Harvard University, where he is a member of the physics department and the School of Engineering and Applied Sciences. He is an experimental physicist who leads a research group studying the properties of the materials encountered in everyday life, like those found in the foods of modernist cuisine, including foams, emulsions, and gels. His group also explores how these materials can be of value to society, from using them to detect pathogens to creating cosmetics.